CONTEMPORARY'S

GED

PREPARATION FOR THE HIGH SCHOOL EQUIVALENCY EXAMINATION

TEST 3:

THE SCIENCE TEST

HARRIET DIAMOND & PHYLLIS DUTWIN

CONSULTANTS

BARBARA CARLSON
GED Instructor
San Juan Adult School
Sacramento, California

DR. ROBERT MITCHELL
Educational Consultant
ABE/GED Instructor
Corvallis, Oregon

CAREN VAN SLYKE
Editorial Director

Contemporary Books, Inc.
Chicago

Library of Congress Cataloging in Publication Data

Diamond, Harriet.
Contemporary's GED.

1. Science—Examinations, questions, etc.
2. General educational development tests. 3. High school equivalency examination. I. Dutwin, Phyllis. II. Contemporary Books, Inc. III. Title. IV. Title: Contemporary's G.E.D. V. Title: GED test 3, the science test. VI. Title: G.E.D. test 3, the science test.
Q182.D53 1985 507'.6 84-27396
ISBN 0-8092-5586-3

Published by Contemporary Books, Inc.
180 North Michigan Avenue, Chicago, Illinois 60601
Manufactured in the United States of America
Library of Congress Catalog Card Number: 84-27396
International Standard Book Number: 0-8092-5586-3

Published simultaneously in Canada by
Beaverbooks, Ltd.
195 Allstate Parkway
Valleywood Business Park
Markham, Ontario L3P 4T8
Canada

Editor
David Caes

Copy Editor
Christine M. Benton

Editorial Assistant
Karin Evans

Contributing Authors
Michaela Komarow, Science Teacher, David Brearley Regional High School, Kenilworth, New Jersey
Rina Zingerman, Biology Teacher, Union Catholic High School, Scotch Plains, New Jersey

Researchers
Gregory Coppa, Science Department, Secondary Schools, East Greenwich, Rhode Island
Thomas Donovan, Science Department, Secondary Schools, East Greenwich, Rhode Island

Production Editor
Kathy Willhoite

Art and Production
Princess Louise El
Ophelia M. Chambliss-Jones
Arvid Carlson

Typography
Terrence Alan Stone

ACKNOWLEDGMENTS

Excerpt on page 88 from "For Your Baby's Sake, Don't Drink!" published by the Department of Alcoholism, Illinois Department of Mental Health.

Article on page 99 by D. Franklin, from *Science News*, April 7, 1984. Reprinted with permission from *Science News*, the weekly newsmagazine of science, copyright 1984 by Science Service, Inc.

Article on page 103 by Rachel Wilder, from *Science Digest*, March 1984. Reprinted by permission of Rachel Wilder.

Article on page 107 from *Science News*, March 3, 1984. Reprinted with permission from *Science News*, the weekly newsmagazine of science, copyright 1984 by Science Service, Inc.

Article on page 151 from *Science News*, December 4, 1982. Reprinted with permission from *Science News*, the weekly newsmagazine of science, copyright 1982 by Science Service, Inc.

Excerpt on page 155 from "Alcohol Alchemy" in *Science Digest*, June 1983. Reprinted by permission from *Science Digest* © 1983 by The Hearst Corporation.

Article on page 167 by Ann Blask, from *Science Digest*, December 1983. Reprinted by permission of Ann Blask.

Excerpt on page 190 from "Tobacco and Your Health: Up in Smoke?" © 1982 Life Skills Education. Reprinted by permission of Life Skills Education, Inc.

Excerpt on page 191 from "Thunderstorms," published by the National Weather Service of the National Oceanic and Atmospheric Administration, NOAA/PA 75009, 1977.

Article on page 192 from *Science News*, March 17, 1984. Reprinted with permission from *Science News*, the weekly newsmagazine of science, copyright 1984 by Science Service, Inc.

Excerpt on page 193 from "And Now a Word about Your Shampoo" by Harold C. Hopkins, from *FDA Consumer*, publication no. (FDA) 76-5002 of the U.S. Department of Health, Education, and Welfare, 1980.

Excerpt on page 194 from "A Nuclear Waste Primer," © 1982 League of Women Voters Education Fund. Used by permission.

Excerpt on page 196 from "Sun, Earth, and Man: The Need to Know," published by the National Aeronautics and Space Administration, U.S. Government Printing Office, 1982.

Table of Contents

INTRODUCING THE SCIENCE TEST — 1

SCIENCE PRE-TEST — 8

SCIENCE READING SKILLS — 22

Introduction 22
Main Idea 22
Locating Supporting Details 25
Organization of Ideas 29
Inferential Understanding 33
Application of Ideas 39
Inferring the Meanings of Unknown Words 46
Answers and Explanations 50

CHARTS, GRAPHS, AND DIAGRAMS — 53

Look at Illustrations Carefully 53
Reading Charts 54
Reading Graphs 57
Reading Diagrams 65
Answers and Explanations 73

BIOLOGY — 75

Introduction to Biology 75
Human Structure and Functions 78
- *Systems of the Human Body* 78
- *Nutrition* 85

Cellular and Molecular Biology 90
- *Cell Structure and Reproduction* 90
- *Genetics* 93
- *Microbes and Microbiology* 98

Evolution and Ecosystems 101
- *The Theory of Evolution* 101

- *Ecology and Ecosystems* 105
- *Flowering Plants* 109

Answers and Explanations 113

EARTH SCIENCE — **118**

Introduction to Earth Science 118
Geology 120
Meteorology and Oceanography 125
Astronomy 129
Answers and Explanations 133

CHEMISTRY — **136**

What Is Chemistry? 136
Atoms and Atomic Theory 139
Compounds and Molecular Theory 144
Chemical Reactions and the Conservation of Matter 149
Organic Chemistry 153
Solutions 156
Answers and Explanations 160

PHYSICS — **163**

Introduction 163
The Laws of Motion 165
States of Matter 169
Light and Sound 172
Electricity and Magnetism 176
Physics and the Future 179
Answers and Explanations 181

SCIENCE POST-TEST — **183**

Introducing The Science Test

Perhaps it has been a long time since you took an important test. Maybe you are not sure about what is required to pass the science test, or you are nervous about the test-taking situation.

This book has been designed to help you succeed on the GED science test. It will provide you with instruction in the skills you need to pass the test, plenty of GED-type practice exercises, and test-taking hints. If you work carefully through this book, you should do well on the science test.

USING THIS BOOK

Contemporary's GED series is a program that you or a teacher can use to determine your individual strengths and weaknesses.

1. Start your work with the pre-test at the beginning of the book. This will help you to preview what the science test includes, but more importantly, it will help you to diagnose the areas you need to concentrate on. Use the skills chart at the end of the pre-test to see what areas need special work.

It would be best for you to work your way carefully through the entire book. However, if your time is limited, and you need more work on another test, you can use the pre-test and skills chart to decide what areas to focus on for the science test.

2. The first two chapters of this book are devoted to specific reading and interpretation skills. To pass the test, you need to understand science passages and answer questions based on them. Also, about 15 percent of the test questions require you to interpret graphs, charts, and diagrams.

The following sections concern the four content areas the test is based on: biology, earth science, physics, and chemistry. Each section has two key features: (1) background essays to familiarize you with science terms and concepts and (2) exercises to give you practice answering questions about passages and pictorial materials.

3. One of the key features of this book is the explanations of answers. If you make a mistake, you can learn from it by reading the explanation that follows the answer and then going back to the question and analyzing your error. The answer key for the content areas also indicates which comprehension skill the question was testing.

4. Finally, the post-test is a simulated GED test. It represents GED-type format and difficulty of questions. It will give you a chance to determine if you are ready for the test and, if not, what areas of the book you still need to review.

THE SCIENCE TEST

There are sixty multiple-choice questions on this test. Forty of them test your ability to interpret science passages, charts, graphs, and diagrams.

Twenty of the test questions are single isolated questions that require an understanding of basic science concepts. For the most part, these questions are based on general principles and the testing of isolated facts is avoided. These discrete questions are the first questions you will see on the GED science test.

THE GED TEST

Q: What does GED stand for?

A: GED stands for the Tests of General Educational Development. The credential received for passing the test is widely recognized by colleges, training schools, and employers as equivalent to a high school diploma. The GED is a national examination developed by the GED Testing Service of the American Council on Education.

The GED tests consist of five examinations in the areas of writing skills, social studies, science, reading skills, and mathematics. While the GED measures skills and knowledge normally acquired in four years of high school, much that you have learned informally or through other types of training can help you pass the test.

The GED test is available in English, French, and Spanish, and on audiocassette, in braille, and in large-print editions.

Q: What should I know to pass the test?

A. The chart below outlines the main content areas, the breakdown of questions, and the time generally allowed per test.

THE GED TEST

Content	Minutes	Number of Questions	Percentage of Test
Test 1 Writing Skills	75	80	
Spelling		10	12.5%
Capitalization & Punctuation		10	12.5%
Grammar & Usage		24	30%
Diction & Style		12	15%
Sentence Structure		12	15%
Logic & Organization		12	15%
Test 2 Social Studies	90	60	
U.S. History		15	25%
Economics		12	20%
Geography		9	15%
Political Science		12	20%
Behavioral Science		12	20%
Test 3 Science	90	60	
Biology		30	50%
Earth Science		12	20%
Chemistry		9	15%
Physics		9	15%
Test 4 Reading Skills	60	40	
Practical Reading		6	15%
General Reading		12	30%
Prose Literature		12	30%
Poetry		5	12.5%
Drama		5	12.5%
Test 5 Mathematics	90	50	
Arithmetic		28	55%
Geometry		10	20%
Algebra		12	25%

On all five tests, you are expected to demonstrate the ability to read and understand what you are reading. You are also tested on many skills you have acquired from life experiences, reading, television, radio, newspapers, contracts, consumer products, and advertising. By and large, you are not expected to recall facts. However, parts of the social studies and science tests require you to remember some information from your past reading and learning.

In addition to the above information, keep these facts in mind:

1. Three of the five tests—reading, science, and social studies—require mainly that you answer questions based on reading or interpreting pictorial materials in these content areas. Developing strong reading and analysis skills is the key to succeeding on these tests.

2. The writing skills test does not require you to write, but rather to detect errors of grammar and usage, punctuation, capitalization, and spelling. Also, you must be able to recognize correct sentence and paragraph construction.

3. The math test consists mainly of word problems to be solved. Therefore, you must be able to combine your ability to perform computations with reading comprehension skills.

Someone once said that an education is what remains after you've forgotten everything else. In many ways, this is what the GED measures.

Q: Can I take the test?

A: Each year, more than 800,000 people take the GED test. In the United States, Canada, and many territories, people who have not graduated from high school and who meet specific eligibility requirements (age, residency, etc.) may take the test. Since eligibility requirements vary, it would be useful to contact your local GED testing center or the director of adult education in your state, province, or territory for specific information.

Q: What is a passing score on the GED?

A: Again, this varies from area to area. To find out what you need to pass the test, contact your local GED testing center. However, you must keep two scores in mind. One score represents the *minimum score* you must get on each test. For example, if your state requires minimum scores of 35, you must get at least 35 points on every test. Additionally, you must meet the requirements of a *minimum average score* on all five tests. For example, if your state requires a minimum average score of 45, you must get a total of 225 points to pass. The two scores together, the minimum score and the minimum average score, determine whether you pass or fail the GED.

To understand this better, look at the scores of three people who took the test in a state that requires a minimum score of 35 and a minimum average score of 45 (225 total). Heino and Bob did not pass, but Maria did. See if you can tell why.

	Heino	Bob	Maria
Test 1	45	38	43
Test 2	47	43	48
Test 3	33	42	47
Test 4	50	40	52
Test 5	50	39	49
	225	202	239

Heino made the total of 225 points but fell below the minimum score on Test 3. Bob passed each test but failed to get the 225 points needed; just passing each test was not enough. Maria passed all the tests and exceeded the minimum average score.

Generally, to receive a GED credential, you must correctly answer half or a little more than half of the questions on each test.

Q: What happens if I don't pass the test?

A: You are allowed to retake some or all of the tests. Again, the number of times that you may retake the tests and the time you must wait before retaking them depend on your state, province, or territory. Some states require you to take a review class or to study on your own for a certain amount of time before retesting.

Q: How can I best prepare for the test?

A: Many libraries, community colleges, adult education centers, churches, and other institutions offer GED preparation classes. Some television stations broadcast classes to prepare people for the test. If you cannot find a GED preparation class locally, contact the director of adult education in your state, province, or territory.

Q: I need to study for the other tests. Are there other materials available?

A. Contemporary Books publishes a wide range of materials to help you prepare for the tests. These books are designed for home study or class use. Contemporary's GED preparation books are available through schools and bookstores and directly from the publisher.

Now let's focus on some useful test-taking tips. As you read this section, you should feel more confident about your ability to succeed on *The Science Test.*

TEST-TAKING TIPS

1. Get prepared physically. Eat a well-balanced meal and get plenty of rest the night before the test so that you will have energy and will be able to think clearly. Last-minute cramming will probably not help as much as a relaxed and rested mind.

2. Arrive early. Be at the testing center at least fifteen to twenty minutes before the starting time. Have time to find the room and to get situated. Keep in mind that many testing centers refuse to admit latecomers.

3. Think positively. Tell yourself you will do well. If you have studied and prepared for the test, you should succeed.

4. Relax during the test. Take a half-minute several times during the test to stretch and breathe deeply, especially if you are feeling anxious or confused.

5. Read the test directions carefully. Be sure you understand how to answer the questions. If you have any questions about the test or about filling in the answer form, ask before the test begins.

6. Know the time limit for each test. The science test has a time limit of ninety minutes.

Some testing centers allow extra time, while others do not. You may be able to find out the policy of your testing center before you take the test, but always work according to the official time limit. If you have extra time, go back and check your answers.

For this sixty-item test, you should allow a maximum of one and a half minutes per question. However, this is not a hard and fast rule. Use it only as a guide to keep yourself within the time limit.

7. Have a strategy for answering questions. You should read through the reading passages or look over the pictorial materials once and then answer the questions that follow. Read each question two or three times to make sure you understand it. It is best to refer back to the passage or illustration in order to confirm your answer choice. Don't try to depend on your memory of what you have just read or seen. Some people like to guide their reading by skimming the questions before reading a passage. Use whichever way works best for you.

8. Don't spend a lot of time on difficult questions. If you're not sure of the answers, go on to the next questions. Answer easier questions first and then go back to the harder questions. If you skip a question, be sure that you have skipped the same number on your answer sheet. Although this is a good strategy for making the most of your time, it can be very easy to get confused and throw off your whole answer key.

Lightly mark the numbers of the questions you did not answer in the margin of your answer sheet so that you know what to go back to. Be sure to erase these marks completely after you answer the questions. In this way, there won't be any confusion when your test is graded.

9. Answer every question on the test. If you're not sure of an answer, take an educated guess. When you leave a question unanswered you will always lose points, but you can possibly gain points if you make a correct guess.

If you must guess, try to eliminate one or more answers that you are sure are not correct. Then, choose from the remaining answers. Remember, you greatly increase your chances if you can eliminate one or two answers before guessing. Of course, guessing should be used only when all else has failed.

10. Clearly fill in the circle for each answer choice. If you erase something, erase it completely. Be sure that you give only one answer per question; otherwise no answer will count.

11. On the science test, the discrete questions come first. The first twenty questions are discrete questions—single, isolated questions that require an understanding of basic science concepts. Many students find these to be the most difficult questions on the GED science test. Don't let them frustrate you. Do your best work and then move into the questions that are based upon your ability to understand science reading passages.

Use the exercises, reviews, and especially the post-test in this book to better understand your test-taking habits and weaknesses. Use them to practice different strategies such as skimming questions first or skipping hard questions until the end. Knowing your own personal test-taking style is important to success on the GED.

SCIENCE PRE-TEST

Introduction

Most people are uncomfortable with tests because they want to do well. It is good that they want to do well, but they make themselves so nervous that they can't do their best work. The best way to deal with tests is to practice.

If tests make you uncomfortable, this pre-test should help you. It is set up like the GED science test so that you can see what the real test will look like, but it has only 40 questions instead of the 60 on the actual GED science test. It will allow you to practice taking a test so that you won't be so nervous when you take the real GED.

The pre-test will also show you your strengths and weaknesses. After you take the pre-test, check your answers and find the areas for which you will need extra preparations.

The pre-test should take you about an hour, but there is no time limit. Some of the questions might be difficult for you, but don't be concerned if you are not sure of the answers. If you do not know an answer, choose the one you think is the best. A good guess is better than no answer at all. In order to receive a passing score on the GED science test you will need to answer only a little more than half of the questions correctly.

The GED science test starts with questions that test your knowledge of science. These are the hardest questions to answer because either you know the answer or you don't. The rest of the test consists of reading passages followed by several questions based on each passage. Don't let problems with the first few knowledge-based questions discourage you. If you learn to do well on the reading passages, you will do well on the actual test.

Answer Grid

1 ① ② ③ ④ ⑤	**14** ① ② ③ ④ ⑤	**27** ① ② ③ ④ ⑤
2 ① ② ③ ④ ⑤	**15** ① ② ③ ④ ⑤	**28** ① ② ③ ④ ⑤
3 ① ② ③ ④ ⑤	**16** ① ② ③ ④ ⑤	**29** ① ② ③ ④ ⑤
4 ① ② ③ ④ ⑤	**17** ① ② ③ ④ ⑤	**30** ① ② ③ ④ ⑤
5 ① ② ③ ④ ⑤	**18** ① ② ③ ④ ⑤	**31** ① ② ③ ④ ⑤
6 ① ② ③ ④ ⑤	**19** ① ② ③ ④ ⑤	**32** ① ② ③ ④ ⑤
7 ① ② ③ ④ ⑤	**20** ① ② ③ ④ ⑤	**33** ① ② ③ ④ ⑤
8 ① ② ③ ④ ⑤	**21** ① ② ③ ④ ⑤	**34** ① ② ③ ④ ⑤
9 ① ② ③ ④ ⑤	**22** ① ② ③ ④ ⑤	**35** ① ② ③ ④ ⑤
10 ① ② ③ ④ ⑤	**23** ① ② ③ ④ ⑤	**36** ① ② ③ ④ ⑤
11 ① ② ③ ④ ⑤	**24** ① ② ③ ④ ⑤	**37** ① ② ③ ④ ⑤
12 ① ② ③ ④ ⑤	**25** ① ② ③ ④ ⑤	**38** ① ② ③ ④ ⑤
13 ① ② ③ ④ ⑤	**26** ① ② ③ ④ ⑤	**39** ① ② ③ ④ ⑤
		40 ① ② ③ ④ ⑤

Directions: Select the best possible answer to each question and fill in the corresponding circle on the answer grid.

1. People cannot live healthily on chocolate cake because
(1) chocolate causes severe skin problems
(2) a continual diet of sweets is always expensive
(3) your body can't break down the cake's carbohydrates
(4) cake does not have all the nutrients necessary for good health
(5) a constant intake of protein will eventually harm your body

2. One disease caused by abnormal cell growth is
(1) cirrhosis
(2) pneumonia
(3) Parkinson's disease
(4) heart disease
(5) cancer

3. Electricity is a moving stream of
(1) clouds
(2) electrons
(3) generators
(4) coal
(5) static

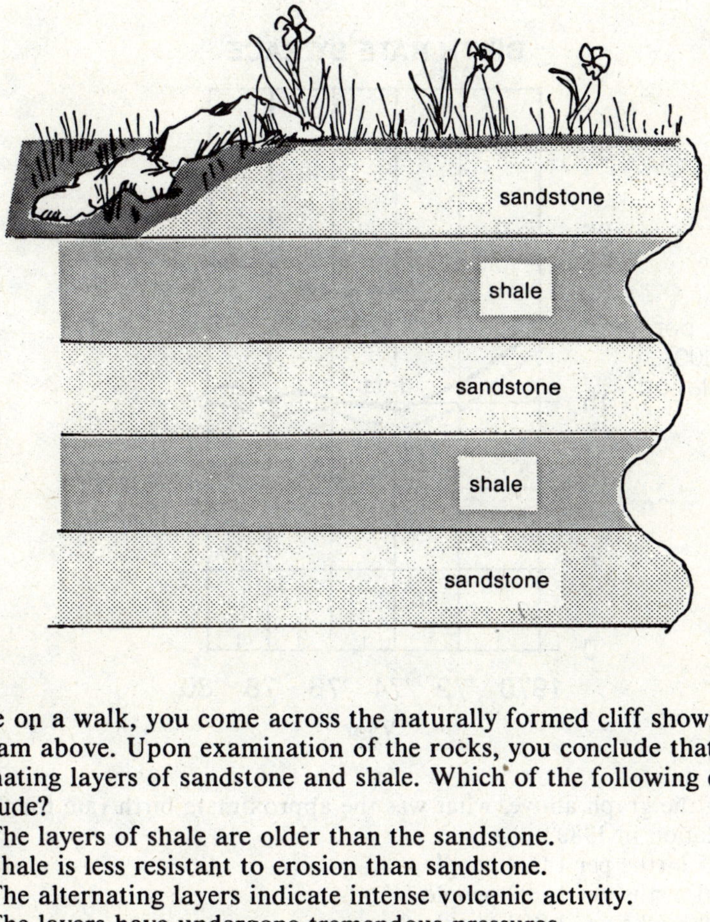

4. While on a walk, you come across the naturally formed cliff shown in the diagram above. Upon examination of the rocks, you conclude that there are alternating layers of sandstone and shale. Which of the following could you conclude?
(1) The layers of shale are older than the sandstone.
(2) Shale is less resistant to erosion than sandstone.
(3) The alternating layers indicate intense volcanic activity.
(4) The layers have undergone tremendous pressures.
(5) The area was once covered by glaciers.

GO ON TO NEXT PAGE.

5. Sand is spread on an icy road to
(1) increase magnetism
(2) decrease magnetism
(3) increase friction
(4) decrease friction
(5) melt the ice

6. If two plates of the earth's crust rub against each other, which of the following is most likely to occur?
(1) An earthquake
(2) A landslide
(3) Soil creep
(4) A tornado
(5) A hurricane

PROPANE

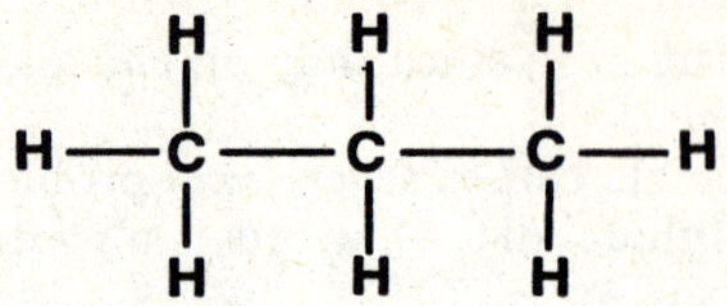

7. The structural formula shown above represents one molecule of propane, a gas. Which of the following is the empirical formula for propane?
(1) CH
(2) 8C3H
(3) 3C8H
(4) C_3H_8
(5) C_8H_3

BIRTH RATE BY RACE

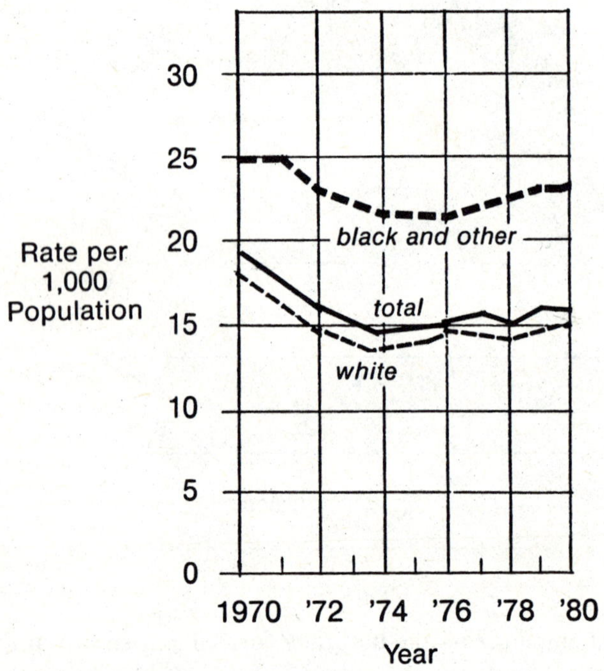

8. Using the graph above, what was the approximate birth rate for the total population in 1980?
(1) 15 births per 1,000 people
(2) 16 births per 1,000 people
(3) 20 births per 1,000 people
(4) 23 births per 1,000 people
(5) 24 births per 1,000 people

GO ON TO NEXT PAGE.

9. Which of the following animals is <u>not</u> a mammal?
(1) A field mouse
(2) A frog
(3) A whale
(4) A rabbit
(5) An elephant

10. How many atoms are in the formula $2H_2O$?
(1) 1
(2) 2
(3) 3
(4) 5
(5) 6

11. In designing roads and bridges, engineers leave spaces and gaps. This is to account for
(1) errors in judgment
(2) the effects of water on the materials
(3) unforeseen changes in the original designs
(4) chemical changes that might be caused by pollution
(5) expansion and contraction due to temperature changes

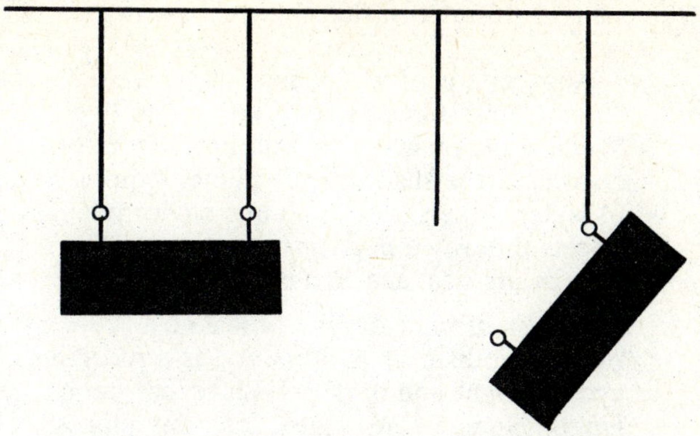

12. The object in the above diagram is suspended from the ceiling by two ropes. If one of the ropes is removed so that the object is suspended by only one rope, which of the following will be true
(1) The remaining rope bears the same load as it did before the other rope was removed.
(2) The load on the remaining rope is increased by 50 percent.
(3) The remaining rope bears twice the load that it did before the other rope was removed.
(4) The load on the remaining rope is increased by at least 400 percent.
(5) None of the above.

GO ON TO NEXT PAGE.

Questions 13–16 refer to the following passage.

All plants need water, light, a suitable climate, and minerals from the soil to live. While some plants grow best in full sun, others do well in the shade. Some plants require lots of water; others less. A plant's environment affects the way a plant grows, its rate of growth, its size, and its ability to reproduce.

One environmental factor is called photoperiodism. A photoperiod is a recurring cycle of light and dark periods of constant length. Some plants, called long-day plants, bloom only if the period of daylight is long. Lettuce is such a plant. Other plants, such as poinsettias and chrysanthemums, are short-day plants, requiring a long, dark period. Still others, such as marigolds, are day-neutral, thriving during both long and short days.

Some plants respond to specific factors in their environment by bending. This behavior is called a tropism. Tropisms take many forms depending on the plant and its environment. Positive phototropism, for example, is the tendency of leaves and stems of a window plant to grow toward the light. Positive geotropism occurs when a plant's roots grow downward toward the earth's center of gravity. When a potted plant is placed on its side, the stem will turn upward within three hours, turning away from the force of gravity. This is called negative geotropism. Roots almost always seek water, exhibiting positive hydrotropism.

13. Growth of plant roots downward toward the center of the earth is called
(1) photoperiodism
(2) negative geotropism
(3) positive geotropism
(4) negative phototropism
(5) positive phototropism

14. Stems and branches that are negatively phototropic will
(1) grow toward the center of the earth
(2) attract light
(3) grow away from a primary light source
(4) grow best in direct sunlight
(5) can absorb large quantities of water

15. A tree near a pond will grow
(1) no roots
(2) roots away from the pond
(3) deep roots
(4) roots toward the sun
(5) roots toward the pond

16. Stems that grow toward the sun show
(1) negative geotropism
(2) negative hydrotropism
(3) positive hydrotropism
(4) negative phototropism
(5) positive phototropism

GO ON TO NEXT PAGE.

Questions 17–21 refer to the following passage.

Mechanical energy is classified in two ways: energy waiting to be used and energy causing action. When an object is at rest at a place from which it can move, such as a rock sitting on top of a hill, it is said to have potential energy (see the diagram). An idling car has this stored energy, as does a ball in one's hand or a cocked pistol. The higher the rock is on the hill, the greater potential energy it has.

The other type of energy is kinetic energy or the energy of motion. Once an object is upset from its position and set into motion—the accelerator of a car is pressed, a ball tossed, or a pistol fired—the potential energy is converted to kinetic energy.

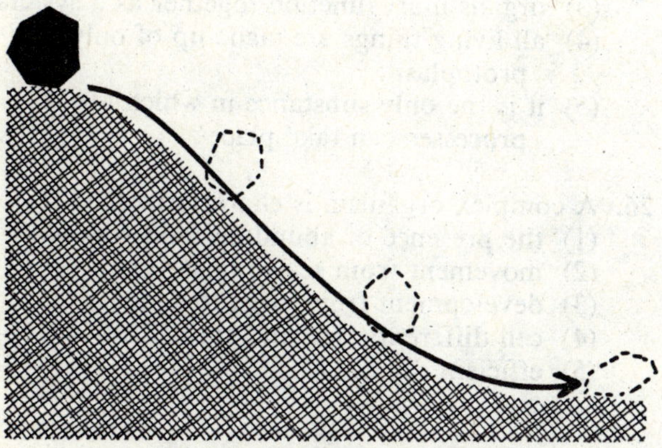

Picture runners poised at the starting line of a race. They represent potential energy, energy waiting to be used. The starting gun is fired, and the runners surge forward. Potential energy has now been converted to kinetic energy—the energy of the runners in motion.

17. The purpose of this selection is to describe
(1) the two types of mechanical energy
(2) how kinetic energy becomes potential energy
(3) the relationship of mass and kinetic energy
(4) how height becomes power
(5) the dissipation of energy

18. Look at the diagram. The rock in the resting position at the top of the hill is said to have
(1) mechanical force
(2) mechanical advantage
(3) no energy
(4) kinetic energy
(5) potential energy

19. The energy of motion is called
(1) kinesthetic energy
(2) fuel-burning energy
(3) kinetic energy
(4) hydraulic energy
(5) potential energy

20. In order for an object to have kinetic energy, there must be
(1) elevation
(2) motion
(3) a storage mechanism
(4) a starting gun
(5) conversion

21. A pendulum at the top of its swing represents
(1) potential energy
(2) kinetic energy
(3) no energy
(4) hydraulic energy
(5) mechanical energy

GO ON TO NEXT PAGE.

Questions 22–26 refer to the following passage.

We now recognize that there are more than 1 million species of animals, which come in an endless variety of sizes, shapes, and structures. However, there are obvious differences in the body structures of animals. The smallest of organisms, a one-celled amoeba, has a very simple digestive process. It simply absorbs food through its cell walls. On the other hand, the more advanced animals such as cats, dogs, and monkeys have very complex digestive processes, with specialized groups of cells that form the different parts of the digestive tract. In these animals, cells are grouped together to form organs that function together as systems.

Despite all of these differences in structure, all living things are made up mostly of protoplasm. Protoplasm is regarded as the only form of matter in which the phenomena of life are manifested, which is why it is called "the basic substance of life."

22. Which of the following would be considered an advanced organism?
(1) The digestive tract
(2) A mouse
(3) An amoeba
(4) A mold
(5) Protoplasm

23. A substance common to all living organisms is
(1) protoplasm
(2) photosynthesis
(3) blood
(4) skeletal systems
(5) wings

24. The main idea of the passage is that
(1) the structures of different species vary from simple to complex, but all are mostly made up of protoplasm
(2) animals must adapt to their environment by developing more complex digestive systems
(3) there are many different kinds of animals
(4) there are more than 1 million species of animals
(5) the level of complexity of an organism is determined by its protoplasm

25. Protoplasm is called "the basic substance of life" because
(1) it enables an organism to develop to a higher form
(2) it aids animals in digesting food
(3) organs must function together as a system
(4) all living things are made up of only protoplasm
(5) it is the only substance in which life processes can take place

26. A complex organism is characterized by
(1) the presence of abundant food supplies
(2) movement from sea to land
(3) development from large to small
(4) cell differentiation and specialization
(5) efficient digestion

GO ON TO NEXT PAGE.

Questions 27–31 refer to the following passage.

Cancer is the term used to describe a group of diseases in which there is an uncontrolled growth of abnormal cells. Masses of these abnormal cells are called tumors. One in six deaths in the United States is caused by cancer, and the number is increasing as people live longer and longer lives. For men, the leading types of cancer are cancers of the lung, prostate, and colon. For women, the leading types of cancer are cancers of the breast, colon, and uterus.

Although scientists have not found out why cancer develops, they have discovered that certain substances play a critical role in causing this disease. For example, tars in tobacco smoke can contribute to lung cancer. Some studies have even shown an increased risk of lung cancer in nonsmokers who are married to smokers. Diets that are high in fats are associated with cancer of the colon. Skin cancer has been associated with prolonged exposure to the ultraviolet rays of the sun through sunbathing. Hundreds of chemicals have been found to cause cancer. Some of these chemicals have entered our food supply as food additives.

Cancer may be cured by removing or destroying the cancerous growth. Currently, surgery, radiation treatment, and chemotherapy are used to destroy cancer cells. Scientists are developing several new treatments that may prove to be effective as cancer cures: the use of interferon, heat therapy, and a holistic approach that involves biofeedback and nutritional counseling.

Research suggests that Americans might have a lower rate of certain cancers if they ate less fatty foods and consumed more high-fiber foods, such as whole grains, fibrous vegetables, and fruits.

27. Which of the following is not mentioned as a possible cause of cancer?
(1) Chemicals in food
(2) Tar in tobacco smoke
(3) Diets high in fats
(4) Exposure to ultraviolet rays of the sun
(5) Lack of exercise

28. Which of the following statements best summarizes the second paragraph?
(1) Scientists have found that a wide variety of substances contribute to cancer.
(2) Scientists have just found a cure for cancer.
(3) Food additives are not a significant factor in the development of cancer.
(4) Tars in tobacco smoke can contribute to lung cancer.
(5) Some food additives have been found to cause cancer.

29. According to the passage, which of the following has been associated with cancer of the colon?
(1) Diets high in fats
(2) Tars in tobacco smoke
(3) Exposure to asbestos
(4) Food additives
(5) Exposure to ultraviolet rays

30. Which of the following most accurately represents the author's opinion about the increasing number of cancer-caused deaths?
(1) The FDA has been lax in notifying the public of possible carcinogenic substances.
(2) The number of chemical additives in our foods is increasing.
(3) As people live longer, they are exposed to more cancer-causing agents.
(4) Since more and more people smoke, the incidence of lung cancer has increased.
(5) More cancer patients are using "quack" cures that only aggravate their conditions and postpone proper treatment.

31. A group of cancer cells is called a
(1) carcinogen
(2) parasite
(3) cirrhosis
(4) colon
(5) tumor

GO ON TO NEXT PAGE.

Questions 32–35 refer to the following passage.

When ocean water is heated by the sun, water from the surface evaporates and becomes an invisible gas known as water vapor, which mixes with the air. As the air moves in currents or winds, the air rises and is cooled.

Cooling decreases the air's ability to hold water so that at a certain temperature, or dew point, the air is saturated and cannot hold any more water vapor. At the dew point, water droplets form. When these become too heavy, gravity takes over, and the particles fall from clouds as rain or snow.

Part of the water that reaches the ground is used by plants and evaporates without ever reaching the sea. Some of the water runs directly into streams, while some sinks to the level of saturated ground. This groundwater and the streams it feeds eventually flow back to the ocean to complete the water cycle.

32. One probable effect of insufficient rainfall is that
(1) streams overflow
(2) evaporation doesn't take place
(3) the air becomes saturated
(4) the ground does not become saturated enough to feed streams
(5) plants continue to release water vapor as usual

33. The term dew point in the second paragraph refers to
(1) the temperature at which condensation takes place
(2) the point at which gravity takes over
(3) water droplets
(4) the temperature at which water saturates the ground
(5) the temperature at which it snows

34. Which of the following pairs of terms has the same meaning?
(1) Dew point and temperature
(2) Evaporated water and water vapor
(3) Cloud and particle
(4) Evaporation and condensation
(5) Precipitation and infiltration

35. The main idea of this passage is that
(1) water travels in an endless cycle of evaporation and condensation
(2) clouds are formed when air is saturated and can hold no more water vapor
(3) oceans are fed through rivers and streams
(4) water condenses when saturated air is cooled to the dew point
(5) plants give off water vapor in a process called photosynthesis

GO ON TO NEXT PAGE.

Questions 36–37 refer to the following passage.

Three hundred miles above the surface of the earth are two circular, magnetic shields or belts. These zones that encircle the earth are called the Van Allen Belts. The zones are named after the physicist James Van Allen, who discovered them in 1958. These belts of intense radiation protect the earth from blasts of dangerous X rays and ultraviolet rays that erupt from the sun.

The Van Allen Belts are held in place by the earth's magnetic field. They also trap incoming high-energy, charged particles and funnel them in at the poles, creating brilliant displays of shimmering lights called auroras in regions near the North and South Poles.

36. What do the Van Allen Belts protect the earth from?
(1) Auroras
(2) Nuclear fallout
(3) X rays and ultraviolet rays
(4) Meteorites
(5) Infrared rays

37. What holds the Van Allen Belts in place?
(1) Intense radiation
(2) The earth's magnetic field
(3) Radiation from the sun
(4) X rays and ultraviolet rays
(5) The auroras

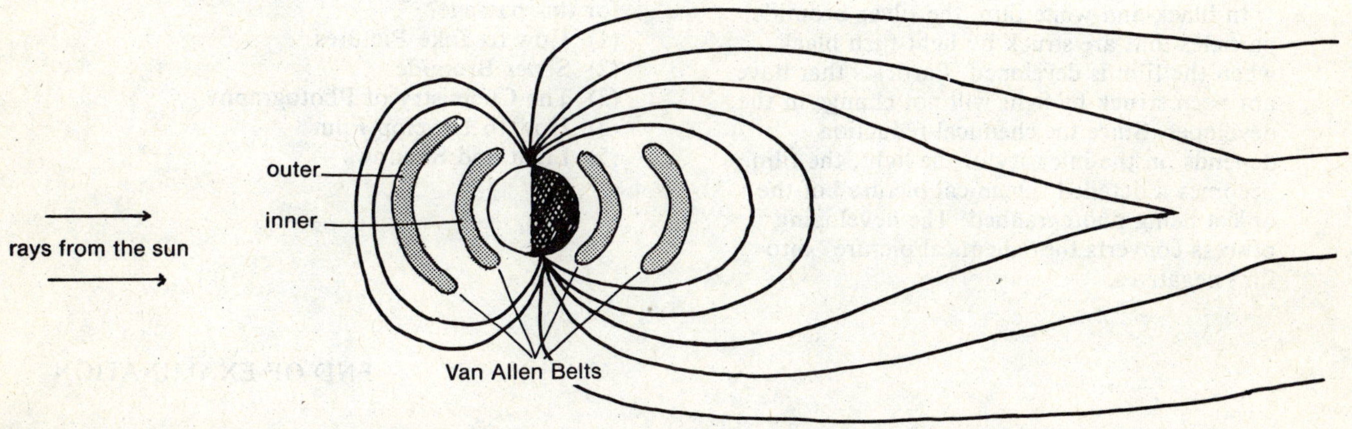

GO ON TO NEXT PAGE.

Questions 38–40 refer to the following passage.

Have you ever wondered how a photograph is made? What actually happens when you push that button to record an image for posterity?

When you push the button on a camera, the shutter opens, allowing light to strike the film in the camera. The film is usually coated with silver bromide. The light rays strike the film and reduce the silver bromide, making it less stable. The chemical reduction looks like this:

$$\underset{\text{silver bromide}}{AgBr} \xrightarrow{\text{light}} \underset{\text{silver}}{Ag} + \underset{\text{bromine}}{Br}$$

In black-and-white film, the silver bromide particles that are struck by light turn black when the film is developed. Particles that have not been struck by light will not change in the developer. Since the chemical reduction depends on the intensity of the light, the film becomes a detailed "chemical picture" of the object being photographed. The developing process converts the "chemical picture" into film negatives.

38. What serves as the agent in reducing silver bromide to silver and bromine?
(1) The camera's shutter
(2) The lens on the camera
(3) Film
(4) Developer
(5) Light

39. What could you conclude if a negative turned out completely black?
(1) The silver and bromine had been resynthesized into silver bromide.
(2) The light rays were not strong enough to reduce the silver bromide.
(3) The film had not been developed.
(4) The film was underexposed to light.
(5) The film was overexposed to light.

40. Which of the following would be a good title for this passage?
(1) How to Take Pictures
(2) Silver Bromide
(3) The Chemistry of Photography
(4) How to Develop Film
(5) Light and Shadows

END OF EXAMINATION

ANSWERS AND EXPLANATIONS

1. (4) A healthy diet is a varied diet that supplies adequate nutrients. Sweets like chocolate cake provide many calories but few nutrients.

2. (5) Cancer is the only choice listed that is caused by abnormal cell growth.

3. (2) An electrical current is a moving stream of electrons.

4. (2) The indentations in the layers of shale indicate that it erodes faster than sandstone. There is no evidence of volcanic activity or other pressures. There is also no evidence of glaciers.

5. (3) Sand on an icy road will increase the friction between a vehicle's tires and the ice. This will make it easier to drive on ice, since the car will have increased traction and will not slide as easily.

6. (1) An earthquake is most likely to occur when crustal plates rub against each other. Landslides may also occur, but these are often triggered by an earthquake.

7. (4) The diagram shows three atoms of carbon and eight of hydrogen. The correct way to write this is C_3H_8.

8. (2) The approximate rate was 16 births per 1,000 people. This does require some skill in estimating, yet you can easily see that the correct answer is greater than 15 yet smaller than 20.

9. (2) A frog is an amphibian, not a mammal. A whale is a mammal. This is a case in which you either know the answer or don't. Take your best guess and go on to the next question.

10. (5) There are four atoms of hydrogen (two times two) and two of oxygen, for a total of six.

11. (5) The effects of temperature change on building materials must be taken into account, especially when you consider the extremes of summer and winter.

12. (3) With two ropes attached, each rope bears half the load. If one of the ropes is removed, the other rope must support the entire load (if it is to stay up). This means the remaining rope must support twice the load that it supported when it shared the load with the other rope.

13. (3) This is mentioned in the third paragraph.

14. (3) Positive phototropism is described as growing toward a light source. Applying this idea you can conclude that *negative phototropism* means "growing away from the light."

15. (5) Using the information on hydrotropisms you could conclude that tree roots would grow in the direction of the source of water.

16. (5) *Phototropism* refers to a plant's growth either toward or away from the sun. A stem with positive phototropism will grow toward light.

17. (1) This entire passage describes the differences and distinctions between the two forms of mechanical energy.

18. (5) The rock is not in motion, so it has potential energy.

19. (3) This detail is stated in the second paragraph.

20. (2) This detail is stated in the second paragraph.

21. (1) A pendulum at the top of its swing is not in motion. It represents potential energy. As it begins its swing, its energy is converted to kinetic energy.

22. (2) A mouse is an animal with a complex digestive process, unlike an amoeba or mold, which are one-celled. Choices (1) and (5) are not organisms.

23. (1) The last paragraph explains that, even though there are vast differences in structure among animals, all living things are made up mostly of protoplasm.

24. (1) Choice (1) brings together the main points of both paragraphs. Choices (2) and (5) are not stated in the passage, and choices (3) and (4) are too specific.

25. (5) The last paragraph states that all living organisms are made up mostly of protoplasm and that it is called "the basic substance of life" because it is regarded as the only form of matter in which life can be manifested.

26. (4) The first paragraph discusses the way in which the cells of complex organisms group together to perform various body functions.

27. (5) The second paragraph mentions several cancer-causing agents. Lack of exercise is not among them.

28. (1) The other options are either too specific or untrue.

29. (1) This fact is stated in the second paragraph.

30. (3) This opinion can be inferred from the first paragraph. The other answers may be true, but they do not reflect the opinion of the author.

31. (5) This fact can be found in the first paragraph.

32. (4) You can conclude that when there is insufficient rainfall, the groundwater will not be sufficient to feed the streams. Answer (2) is incorrect because evaporation takes place as long as there is water present.

33. (1) From the second paragraph you can conclude that the dew point is the temperature at which condensation, or the formation of water droplets, takes place in saturated air.

34. (2) The passage uses *evaporation* and *water vapor* to refer to the same process.

35. (1) While choices (2), (3), and (4) are all true, they are too specific to be the main idea. Answer (5) is not discussed.

36. (3) This fact is stated in the first paragraph.

37. (2) This fact is stated in the second paragraph.

38. (5) Light reduces silver bromide to silver and bromine. Look at the chemical equation.

39. (5) The last paragraph states that reduced particles of silver bromide turn black when film is developed. If a negative is completely black, then all of the silver bromide must have been reduced by light, indicating too much light, or overexposure.

40. (3) This is the only title that summarizes this passage. The other titles concern a specific topic discussed in the passage.

PRE-TEST EVALUATION CHART

On the chart below, circle the numbers of the questions you missed.

	Main Idea	Finding Details	Application of Ideas	Inferences and Conclusions	Prior Knowledge
Biology	24, 28	***8***, 13, 23, 27, 29, 31	14, 16	15, 22, 25, 26, 30	1, 2, 9
Earth Science	35	36, 37	32	***4***, 33, 34	6
Chemistry	40	38	39	***7***	10
Physics	17	19, 20	***12***, 21	***18***	3, 5, 11

If you got more than 20 questions correct you are in pretty good shape.

You probably did better in some areas than you did in others. That is normal. If you didn't do well on the pre-test, you probably will want to concentrate first on the Science Reading Skills section of this book. This will help you with the reading skills that are useful on the GED science test.

There may be areas of science in which you need to do special work. If you got half or more than half of the questions wrong in any one area, you will probably want to spend more time going through the material on that area than on others.

On the pre-test there were eight questions that tested prior knowledge. There is no way to guarantee that you are prepared for any possible prior knowledge question that you will be asked. Each section of this book is constructed to give you an overview of scientific terms and concepts, so at least you will have a general knowledge of each content area.

As you noticed on the pre-test, there are more questions in the Biology section than in the other areas. Half of the questions on the GED science test will be on Biology. If you have a very limited amount of time, you would do best to focus your efforts on the Reading and Biology sections.

The numbers in darker type indicate questions based on graphic or pictorial materials. If you miss more than two of these, you should pay close attention to the chapter that starts on page 53.

Science Reading Skills

INTRODUCTION

Most of the questions you will see on the GED science test involve interpreting short reading sections called "passages." The key to passing the science test, especially for those of you who do not have a strong science background, is learning to understand these short science passages.

The first section of this GED science book will help you develop specific reading skills and will demonstrate how you can use these skills in scientific material. For each of the separate skills, you will "walk through" several examples; then you will have a chance to try each skill on your own.

MAIN IDEA

Did you know that there is a **main idea** in almost everything you read? In fact, quickly finding the main idea of a passage is one of the most important reading skills used on the GED. Your understanding of what you read depends greatly on this skill. The main idea is important because it is a short summary of what the passage is about.

Many people believe that the main idea is always the first sentence in a passage. This belief is not correct. In fact, the main idea can appear anywhere within the passage. It can be the first sentence, the last sentence, or a sentence in the middle. What's more, the main idea may not be stated as a single sentence at all! In this case, the main idea is simply a summary that you make in your mind of the information presented in the passage.

The above ideas will become clear as you read the examples in this section. Read the following paragraph. Which sentence states the main idea? ______

EXAMPLE

[1]The presence of living things brings about various changes in the environment. [2]For example, when beavers dam a stream, many changes may occur. [3]Before long, land above the dam is flooded. [4]Trees, bushes, and plant life that once grew in the now-flooded area die. [5]They are replaced by plants that are able to live in water.

The first sentence makes a general statement about living things bringing about changes in the environment. Sentences 2, 3, 4, and 5 state that floods, death of vegetation, and growth of new vegetation occur when beavers dam a stream. Only one sentence in the passage is general enough to include the ideas of the others. That is sentence 1. In this passage, the first sentence states the main idea.

Try another paragraph. Remember that the main idea can occur anywhere in the passage. Which sentence states the main idea? ______

EXAMPLE

[1]Did you know that parents catch most of their colds from their children? [2]Young children between two and six years old usually have twice as many colds as nine-year-old children. [3]Nine-year-olds usually have twice as many colds as twelve-year-olds. [4]Generally speaking, as we get older, we get fewer colds.

Finding this main idea is a little more difficult than in the previous passage. The main idea is stated in the last sentence. Sentences 2 and 3 provide the details to show that we get fewer colds as we get older. The first sentence can easily fool you, but it does not contain the main idea. It introduces and supports the main idea.

SCIENCE READING EXERCISE 1

Directions: Read the following passages and answer the questions. Fill in the circle that corresponds to the correct answer.

The earth as a source of energy has a major rival—the sun. The sun offers many advantages over geothermal energy. One is that, as a source, the sun is unlimited. Another advantage of solar energy is that it is pollution-free. Solar energy could help to alleviate the energy crisis if it could be converted into electricity and stored on a large scale.

The sun's energy has been used in a limited way to heat and distill water, to cook food, and to heat homes. Also, solar cells that transform sunlight directly into electric current have been used by the space program as a reliable source of electricity. Unfortunately, the devices used to accomplish these tasks are far too expensive for extensive use at this time.

1. The purpose of this passage is to describe — 1 ① ② ③ ④ ⑤
(1) the advantages and uses of solar energy
(2) the devices used to harness solar energy
(3) the disadvantages of solar energy
(4) converting solar energy into electricity
(5) geothermal energy

2. Which of the following would make the best title for the passage? — 2 ① ② ③ ④ ⑤
(1) Pollution-free Energy
(2) The Sun
(3) Unlimited Energy
(4) Solar Energy Could Meet Today's Needs
(5) Solar Energy in the Space Program

Woods Hole is a resort area near Falmouth, Massachusetts. Bicycling, swimming, and sailing are among the activities that both relax and invigorate visitors.

A variety of scientists also congregate at this vacation village, but their activities are quite different from the usual sun and surf variety. Geologists dive to the ocean floor to examine underwater mountains and valleys. Oceanographers test the ocean currents. Biologists study the effects of ocean pollution and search for new sources of seafood. Chemists at Woods Hole study oil spills and other pollutants in the marine food cycle and learn about the chemical interactions among the creatures of the sea.

3. The best title for this passage is: — 3 ① ② ③ ④ ⑤
(1) Mysteries of the Deep
(2) An Exciting Summer Vacation
(3) Hardworking Scientists
(4) Geologists at Woods Hole
(5) Research Activities at Woods Hole

4. The passage emphasizes that, although Woods Hole is a summer resort, — 4 ① ② ③ ④ ⑤
(1) a great deal of significant scientific study goes on there
(2) too much scientific activity goes on there
(3) people shouldn't vacation there because scientists are polluting the water
(4) many scientists escape from their work and go to Woods Hole
(5) the effects of ocean pollution make swimming unpleasant

Air pollution can be seen and felt in the smog that hovers over our major cities. Pollutants burn our eyes, irritate our lungs, and blot out sunlight. The incomplete combustion of gasoline releases carbon monoxide (CO) and hydrocarbons into the atmosphere. Since CO reduces the amount of oxygen supplied to the body's cells, it is particularly dangerous to people with heart or lung disease. Because of the health hazards posed by air pollution, it is easier to understand the increasing emphasis on stringent measures to protect the environment.

5. Which of the following is the main idea of this passage? **5** ① ② ③ ④ ⑤

(1) Air pollution can be seen in the smog over our cities.
(2) Air pollution burns our eyes and irritates our lungs.
(3) High levels of CO are dangerous to people with heart and lung disease.
(4) Air pollution is a health hazard.
(5) We must protect our environment.

6. The passage suggests that, even though air pollution is a problem, **6** ① ② ③ ④ ⑤

(1) smog will continue to increase
(2) government regulation can help us preserve our environment
(3) it is not a serious health hazard
(4) it is especially dangerous to people with heart or lung disease
(5) the situation will only get worse

For answers and explanations, see page 50.

LOCATING SUPPORTING DETAILS

You have practiced a skill that is central to reading: finding the main idea. Another important reading skill is locating **supporting details**.

You have already learned that the main idea is supported by the details in a passage. On the GED test, you will be asked to locate some of these details. You should be aware that, when you are looking for supporting details, you do not have to reread an entire selection in order to find them. Instead, you may **scan** the material.

When you scan for an answer, you run your eyes quickly over the passage, looking for the detail or part that relates to the question. This can be done easily, especially if the question involves a number or a date. Let's walk through two examples to see how this can be done.

EXAMPLE

Scientists have been studying the adverse physical effects of long-term space travel. Muscles may atrophy as a result of prolonged weightlessness. Bones can lose .5 percent of their calcium monthly. The fatigue from space flight causes sleep problems. Pulse rates have been found to increase because the heart shrinks by about 10 percent.

Which of the following is not mentioned as an adverse effect of long-term space travel?
(1) Sleep problems
(2) Acute dizziness
(3) Increased pulse rate
(4) Bones lose calcium
(5) Muscles atrophy

After you read the question, you should go back to the passage to look for the adverse effects of space travel. By scanning, you can locate the information you need. With this particular type of question it is important that you go back to check your recollection of the information. Choice (2) was not mentioned by the author.

That example was relatively easy. Let's try another one.

EXAMPLE

[1]The processes that break down stone into smaller pieces of matter are influenced by several factors. [2]Time is an important factor in determining the extent of decay. [3]Climate is another influence since stone will decay faster in a moist climate than in dry air. [4]Stone that is buried beneath the soil will decay more rapidly because acids are working on it.

How can climate affect the decay of stone?
(1) Acids in the atmosphere decay stone.
(2) Stone breaks down into smaller pieces.
(3) Stone will break down faster in a dry climate.
(4) It does not take much time for stone to break down.
(5) Stone deteriorates faster in moist climates.

Since the question asks about the effects of climate, it would be best to go back to the passage to read exactly what the passage says. By comparing sentence 3 with the possible answers to this question, you can quickly identify the correct response, choice (5).

The next passage contains several words that you may not know. On the GED test you may find scientific words you have never seen before. To do the test well, you do not have to know all the terms. You should be able to use your reading skills to find the correct answer.

EXAMPLE

Loudness depends upon two factors: frequency and the intensity level of a sound. Low-frequency sounds (below 900 hertz) are thought to be softer than high-frequency sounds (between 900 and 5,000 hertz)—if the intensity remains the same. A sound level meter is designed so that tones or narrow bands of noise will all sound the same if their sound levels are about 40 decibels.

900 hertz is thought to be
(1) the density level of sound
(2) about 40 decibels
(3) the lower limit for high-frequency sounds
(4) a typical low-frequency sound
(5) too loud for a cat to hear

Loudness depends upon
(1) tones or narrow bands of noise
(2) frequency and the intensity level of a sound
(3) low-frequency sounds
(4) your hearing
(5) high-frequency sounds

The answer to the first example question is (3). You probably looked first for the number *900*. The second step was to note the context in which the number was written. The number *900* appears in two places in the passage, but only in the second instance can you find the answer to the question. Since high-frequency sounds are between 900 and 5,000 hertz, 900 hertz would be the lower limit for these sounds.

The correct answer to the second example question is (2). The words in this particular question were taken directly from the passage. If you don't recall these words from your first reading, scan to find them. "Loudness depends upon two factors. . . ." Read further to see which of the choices best completes the sentence. When you find the words "frequency and intensity level," you know that answer (2) is correct.

Now you are going to have a chance to locate supporting details. Read the following passages. When you finish, read each question and then scan for the correct answer.

SCIENCE READING EXERCISE 2

Directions: Read the passages and answer the questions that follow. Fill in the circle that corresponds to the correct answer.

A person can get dizzy examining the causes of dizziness. Dizziness can be caused by disorders of the circulatory system, including high blood pressure, stroke, anemia, and leukemia. Tumors, hemorrhages, and epilepsy also may cause dizziness. Dizziness is the mildest symptom of a very serious disorder called Ménière's disease. Certain infectious diseases can cause temporary dizziness. Causes of less serious dizzy spells could include excessive earwax or new eyeglasses.

7. It is clear from this passage that dizziness can be
(1) caused by a limited number of factors
(2) induced only by blood or vascular problems
(3) related to one or two causes
(4) developed only from infectious diseases
(5) caused by any of a broad range of factors

7 ① ② ③ ④ ⑤

8. Some of the less worrisome causes of of dizziness are
(1) high blood pressure and strokes
(2) excessive ear wax and new eyeglasses
(3) anemia and leukemia
(4) strokes and leukemia
(5) infectious diseases and Ménière's disease

8 ① ② ③ ④ ⑤

The Environmental Protection Agency (EPA) is investigating Times Beach, Missouri, in addition to 132 other sites in that state that may have been contaminated by the chemical dioxin. Technicians have discovered dioxin at levels as high as 100 parts per billion along roadsides in Times Beach. Prior to this, the EPA had ordered the cleanup of sites containing only 1 part per billion.

The major focus of the current effort is to prevent public exposure to the dangerous chemical, which is in the ground as well as inside some homes. It is feared that dioxin will seep through the soil into the groundwater. Fortunately, dioxin is one of the few chemicals that bonds itself to soil, minimizing its chances of seeping into the water supplies.

9. Because dioxin bonds itself to soil,
(1) it is not dangerous
(2) the EPA is investigating it
(3) it will definitely contaminate the water system
(4) it may not seep into the water system
(5) it is in some homes

9 ① ② ③ ④ ⑤

10. Prior to the discovery of dioxin at Times Beach, what was the level of dioxin that would trigger an EPA order to clean up a site?

(1) 100 parts per billion
(2) 50 parts per billion
(3) 1 part per billion
(4) 1 part per million
(5) less than 1 part per million

10 ① ② ③ ④ ⑤

For answers and explanations, see page 50.

ORGANIZATION OF IDEAS

There are different ways that supporting details are organized in science reading material. Two methods of science writing involve sequence of events and cause-and-effect relationships.

Sequence of Events

In reading a description of a scientific process or of a natural phenomenon, you will notice that the details are presented in a logical order. Here is an example:

EXAMPLE

[1]Heat pasteurized milk to 39°C (102°F). [2]Salt the milled curd. [3]Press salted, milled curd into hoops for shaping. [4]After three to four days, drop formed cheese into wax. [5]For milder cheddar, cold-cure at 0°–5°C (22°–41°F).

As you read the above passage, you probably noticed that each event in the **sequence** logically followed the one preceding it. The curing process would not make any sense if sentences 2 and 3 were reversed.

Read the following passage and answer the questions that follow it. This is another passage with technical language. Don't let it throw you. Use your reading skills to work through it.

EXAMPLE

[1]The incubation period for typhoid fever lasts eight to fourteen days. [2]The symptoms of the disease begin as the bacilli enter the bloodstream. [3]During the incubation period, the typhoid bacilli travel through the alimentary canal to the small intestine, where they settle in lymphatic tissues known as Peyer's patches. [4]From Peyer's patches, the bacilli enter the lymph system and are transported into the bloodstream.

Where do the typhoid bacilli lodge as soon as they have passed into the small intestine?
(1) Typhoid fever
(2) Small intestine
(3) Alimentary canal
(4) Bloodstream
(5) Peyer's patches

Where are the typhoid bacilli before they enter the small intestine?
(1) Alimentary canal
(2) Small intestine
(3) Typhoid fever
(4) Lymphatic tissues
(5) Peyer's patches

The correct answer to the first question is (5). The third sentence states that the typhoid bacilli travel to the small intestine and settle in the Peyer's patches.

The correct answer to the second question is (1). The typhoid bacilli are in the alimentary canal before entering the small intestine. Choices (4) and (5) are places in the small intestine. Choice (3) doesn't make sense.

Cause-and-Effect Relationships

In your science reading, you will note that certain actions (causes) lead to specific results (effects). These are known as **cause-and-effect** relationships. Often, a cause will produce a certain effect, which will become a cause of something else. For example, a lack of B vitamins in the diet may cause eczema. Eczema is the result (effect). Eczema, in turn, becomes the cause of scaling and itching.

Now read the following paragraph and answer the questions that follow.

EXAMPLE

Toxoplasmosis, an infectious disease of animals and people, is caused by a parasitic protozoan. The disease is commonly transmitted by the household cat. Once acquired, toxoplasmosis is usually asymptomatic (without symptoms), although it may produce mild symptoms that resemble those of mononucleosis or a cold. The disease is difficult to recognize.

According to the passage, a parasitic protozoan
(1) always causes mononucleosis in cats
(2) causes toxoplasmosis in cats only
(3) causes toxoplasmosis in animals and people
(4) causes colds in animals and people
(5) is acquired asymptomatically

If toxoplasmosis is the effect, what is the cause?
(1) The common cold
(2) Humans
(3) Cats
(4) A parasitic protozoan
(5) Mononucleosis

The correct answer to the first question is (3). A cause-and-effect relationship between a parasitic protozoan and toxoplasmosis is described in the passage. In the first sentence, toxoplasmosis is introduced as a disease that affects both animals and people.

The correct answer to the second question is (4). The first sentence states that toxoplasmosis is caused by a parasitic protozoan.

SCIENCE READING EXERCISE 3

Directions: Read the passages and answer the questions that follow. Fill in the circle that corresponds to the correct answer.

Osteoporosis is a degenerative disease associated with aging. It causes the bones to become porous, brittle, and susceptible to breaking. When the spine is involved, the vertebrae may collapse, causing curvature of the spine, backache, and a decrease in height.

Osteoporosis is usually associated with a lack of calcium. The blood seeks to maintain normal calcium levels. In periods of deprivation, calcium is removed from the skeleton.

Osteoporosis is especially common among women after menopause. It is also seen in men, especially chronic alcoholics. The cause and remedy for osteoporosis are not known. It seems that the right amount of calcium-rich products, the lifelong use of fluoridated water, and regular activity decrease the risk of osteoporosis in later years.

11. What is the cause of osteoporosis?
(1) Calcium-rich foods
(2) Regular activity
(3) Curvature
(4) Overabundance of calcium
(5) The cause is not known

11 ① ② ③ ④ ⑤

12. Which of the following may happen with the onset of osteoporosis?
(1) The bones become brittle.
(2) The spine is broken.
(3) Alcoholism increases.
(4) Women experience menopause.
(5) The blood is depleted of calcium.

12 ① ② ③ ④ ⑤

13. Which of the following may prove to decrease the risk of osteoporosis?
(1) Corrective surgery
(2) Calcium-rich products
(3) High levels of alcohol
(4) Aspirin
(5) Lack of exercise

13 ① ② ③ ④ ⑤

In nature, the animal population is limited by certain factors. The presence of predators and the amount of available food are two critical factors. Disease-causing organisms also limit population growth. Although most species of animals produce large numbers of offspring, relatively few reach adulthood. Plant and animal populations tend to be stable from year to year.

14. An unusually large increase in the number of predators or disease-causing organisms would
(1) cause the prey animal population to grow
(2) provide more than enough food for the population
(3) cause the prey animal population to decrease
(4) not affect the prey animal population
(5) allow large numbers of offspring to reach adulthood

14 ① ② ③ ④ ⑤

15. If humans eliminated the predators, how would this initially affect the number of prey animals?
(1) The number would remain constant.
(2) The number would slowly decrease.
(3) The number would rapidly decrease.
(4) The number would increase.
(5) The number would be subject to minor fluctuations.

15 ① ② ③ ④ ⑤

For answers and explanations, see page 50.

INFERENTIAL UNDERSTANDING

So far we have worked with reading skills that require you to locate information in a reading passage. There will also be questions on the GED science test that require you to draw conclusions and make inferences.

Drawing Conclusions

When you read a question that requires you to draw a conclusion, you will not find the answer stated directly in the passage. You use the information given in a passage to predict an outcome or to draw a reasonable conclusion based on the facts you know. For example, if you know that water is a conductor of electricity, and electricity is harmful—if not deadly—to a human being, you could conclude that it is very dangerous to operate an electrical appliance with wet hands or to leave an electrical appliance near a bathtub.

Let's go ahead and practice drawing conclusions. Read the following passage and answer the question that follows.

EXAMPLE

Fossils reveal many interesting things to paleontologists. These scientists are able to determine the age of rocks by the fossils in them. Fossils also tell whether the rocks were formed on the land or under the ocean. Most rocks that contain animal or land plant fossils originated on land. Those that contain marine shell fossils were formed under the ocean.

You can conclude that the study of fossils gives paleontologists

(1) the exact location of deep ocean trenches
(2) the ability to date plant fossils more accurately than animal fossils
(3) information about the kinds of life that existed during different periods of the earth's history
(4) many kinds of information, but nothing that tells them the ages of rocks
(5) the information they need to explain the origin of the earth

The correct answer is (3). The second sentence states that scientists can date fossils and rocks. It is logical to conclude, based on this fact, that scientists can tell what life existed at different stages in the earth's history. There is not enough information to draw the conclusions offered in choices (1) and (2). Choice (4) is not correct since the second sentence states that paleontologists can determine the age of rocks. While it is true that fossils preserve the history of the earth, this passage gives insufficient information to draw the conclusion in choice (5).

SCIENCE READING EXERCISE 4

Directions: Read the following passages and answer the questions. Fill in the circle that corresponds to the correct answer.

Almost all living organisms inhabit areas where temperatures range from 0° F to 110° F. Most organisms' metabolic activity and behavior have developed in response to temperature. The body temperature of living organisms, with the exception of birds and mammals, is influenced primarily by the external environment.

16. Which of the following statements is the best conclusion based on this passage?
(1) A bird would die only if the temperature were above 114° F.
(2) The optimum temperature for all mammals is 56° F.
(3) The body temperature of all plants and animals is influenced mainly by external factors
(4) The body temperature of birds and mammals is influenced by factors other than external environment.
(5) The body temperature of all plants and animals, except mammals and birds, is determined only by external factors.

16 ① ② ③ ④ ⑤

17. The main idea of this passage is that
(1) temperature is significantly related to important life functions of plants and animals
(2) almost all life exists within a temperature range of 0°–123° F
(3) Centigrade temperature does not affect living things
(4) birds and reptiles cannot survive in the samè environment
(5) normal human body temperature is 98.6° F

17 ① ② ③ ④ ⑤

18. Which of the following organisms would be most affected by a change in the air temperature?
(1) Bird
(2) Lizard
(3) Monkey
(4) Horse
(5) Human being

18 ① ② ③ ④ ⑤

When an egg is fertilized, it undergoes cell division to form an embryo. The protection of the embryo varies among different species. In most lower forms of life, the embryo is encased in a protective shell and discharged from the female's body. Some animals, such as birds, lay their eggs before their young are fully developed. Others hatch the eggs within the mother's body. In the higher-level animals, the embryo grows in the uterus and is nourished by the mother until it is developed sufficiently for birth. In plants, the fertilized ovule develops into a seed that contains an embryo, a store of food, and a protective covering.

19. If an organism nurtures an embryo in its uterus until the embryo is mature enough for birth, how would this organism be categorized by the author of this passage?

(1) Bird
(2) Plant
(3) Lower-level animal
(4) Higher-level animal
(5) None of the above

19 ① ② ③ ④ ⑤

20. You could conclude that an injury to the covering of the embryo would result in

(1) fertilization
(2) nourishment
(3) a speeding up of birth
(4) damage to the embryo
(5) damage to the mother

20 ① ② ③ ④ ⑤

For answers and explanations, see page 51.

Making Inferences

When you make an inference, you make an educated guess. Some people refer to this as "reading between the lines." Before we actually jump in and use this skill, let's look at a few examples to see how it is actually used.

EXAMPLE

Any deviation from the desired width or weight of a given material can be detected immediately by a radiation thickness gauge, a device that is used to measure coatings on metals and other materials. The detected deviation is relayed to an indicator that alerts the production foreman to the need for an adjustment. Often, the gauges are hooked up to electronic computers and other mechanisms that automatically make the required adjustments.

From this passage, we can infer that
(1) the radiation thickness is often connected to a manually operated computer
(2) the radiation is a computer
(3) the radiation thickness is used to cut materials down to a desired width or weight
(4) the radiation thickness gauge dispenses with the need for a production foreman
(5) the use of a radiation thickness gauge leads to the production of a highly uniform product

The right answer is (5). Since the gauge can detect the slightest deviation in width or weight, its use can lead to the production of a highly uniform product, one whose coating always has the same width and weight. Choices (1) and (4) contain incorrect information. Choices (2) and (3) are incorrect inferences, since the gauge is separate from the computer (2), and the gauge only detects deviation in width and weight and does not do the actual cutting (3).

Making inferences is different from the types of questions that require you to find a stated main idea or detail. You must draw on information found in the passage to make an educated guess or interpretation.

Now try another short passage and answer the question by making an inference.

EXAMPLE

Until the sixteenth century, most scientists thought they were observing the sun revolving around the earth. Astronomers accepted this theory, which had been formulated 1,400 years earlier by Ptolemy. According to Ptolemy, the earth was the center of the universe. He believed that the earth was stationary and that the sun revolved around it.

From this passage, we can infer that Ptolemy's theory
(1) was not proven correct until the sixteenth century
(2) remained the only explanation until the twentieth century
(3) was not believed by anyone else
(4) incorrectly explained the orbiting of the sun and other bodies
(5) laid the groundwork for the space probes of the twentieth century

The correct answer is (4). Ptolemy thought that the earth was stationary and that the movement of the other heavenly bodies proved his theory. What he did not realize was that the earth was also in motion. Choices (1), (2), and (3) are incorrect, based upon the information found in the first two sentences. Choice (5) is also incorrect. It is not likely that an incorrect theory would lay the groundwork for twentieth-century space probes.

SCIENCE READING EXERCISE 5

Directions: Read the following passages and answer the questions. Fill in the circle that corresponds to the correct answer.

Aluminum is the most abundant metal in the earth's crust. It is light in weight, making it important to the building, automobile, and aircraft industries. Aluminum is capable of conducting both heat and electricity. One of aluminum's most important characteristics is its resistance to corrosion. While other metals are affected adversely by oxygen, aluminum is not. It reacts quickly with oxygen, combining with it to form a tough surface film that prevents further corrosive action.

21. Aluminum is important to the automotive and aircraft industries because it — 21 ① ② ③ ④ ⑤

(1) prevents serious accidents on the ground and in the air
(2) does not conduct electricity
(3) promotes corrosion in automobiles
(4) resists corrosion
(5) makes automobiles better looking and easier to paint

22. It is reasonable to assume that aluminum is an unusual metal because of its — 22 ① ② ③ ④ ⑤

(1) use in manufacturing cars
(2) resistance to corrosion
(3) light weight
(4) adverse reaction to oxygen
(5) tough film

The most serious problem an alcoholic faces is a diseased liver. The liver is the key organ for processing alcohol, as well as other harmful substances, out of the blood. With excessive and prolonged use of alcohol, the liver suffers severe toxic effects and eventual damage to its tissues. The condition that finally renders the liver ineffective is called *cirrhosis*.

The alcoholic's liver is scarred. Groups of hard cells replace normal spongy tissue. This prevents the absorption and processing of substances in the blood. Unable to function properly, the liver cannot perform its life-sustaining work. Alcoholics with liver problems must give up drinking or risk the life-threatening consequences of cirrhosis. If liver damage is not too severe, a liver can repair itself in the absence of alcohol.

An alcoholic may also suffer nutritional deficiencies. Since alcohol has many calories, the alcoholic tends to get calories by drinking. Alcohol lacks vitamins, minerals, and other essential nutrients, which seems to account for the malnourished condition of so many alcoholics.

23. Which of the following is a consequence of liver damage?

(1) Loss of appetite
(2) Brain damage
(3) Excessive drinking
(4) Insufficient absorption of nutrients
(5) Spongy liver tissue

23 ① ② ③ ④ ⑤

24. Why are many alcoholics malnourished?

(1) Alcohol accounts for most of their caloric intake, but it has few essential nutrients.
(2) A diseased liver allows only alcohol to enter the bloodstream.
(3) The hard cells in the liver are replaced by spongy cells.
(4) Nutrients are absorbed by the blood.
(5) The liver removes all the nutrients from the bloodstream.

24 ① ② ③ ④ ⑤

Although it appears solid, the earth's crust is constantly moving. Under the weight of mountain ranges and continents, rocks creep and flow. The shift of weight causes strains and breakages in the earth's crust.

The process of change in the crust may take millions of years, or it may occur rapidly. Faults may be created beneath the surface, and sudden shifts can cause earthquakes.

A slower process, called *folding*, may also occur. An outstanding example of this is in Scandinavia, where coastlines have risen above sea level. The rocks that were originally deposited in horizontal layers have been warped or bent upward and downward by crustal movement.

Mountains may rise not only from faulting or folding but also from volcanic activity. Examples of this are Mount St. Helens and the Hawaiian Islands.

25. Which of the following can be inferred from the passage?
(1) The face of the earth is constantly, although slowly, changing.
(2) All mountains are contantly rising.
(3) The earth's crust is stable.
(4) Faulting always results in cliffs.
(5) Warping occurs only in an upward direction.

25 ① ② ③ ④ ⑤

26. Scandinavia, Mount St. Helens, and the islands of Hawaii are
(1) the only areas whose landforms result from crustal strain
(2) the only areas where faulting occurs
(3) three areas where volcanoes exist
(4) landforms resulting from crustal strain
(5) stable, unchanging landforms

26 ① ② ③ ④ ⑤

For answers and explanations, see page 51.

APPLICATION OF IDEAS

On the GED science test, you may encounter some questions that require you to apply science principles. When you **apply an idea** from a science reading passage, you may take information or an idea from the passage and apply it to a similar set of circumstances that is not mentioned in the passage. You may have to apply a scientific principle or analyze a chemical formula. In the following sections, you will learn how to use and apply each of these kinds of information.

Applying Scientific Principles

You can apply general scientific principles to situations described in a science passage. You can use information that you already have to help you understand a new situation. For instance, you already know that water freezes at 32°F. Use this information to answer the following question.

EXAMPLE

Snow fell heavily Saturday morning. During the afternoon, temperatures warmed slightly and the snow changed to sleet, a mixture of snow and rain, which melted some of the snow. In the evening, the temperature dropped well below 30°F.

Driving conditions were
(1) not dangerous because the snow had melted
(2) difficult because the roads were snow-covered
(3) dangerous because the sleet and melted snow had turned to ice
(4) unpleasant because the streets were flooded
(5) not unusual

Answer (3) is correct. Once the evening temperature dropped below 30° F, the sleet and melted snow froze. The highways were probably glazed over with ice. To answer this question correctly, you had to apply the knowledge that water freezes at 32° F.

Now, try another example.

EXAMPLE

Human metabolism generally slows down with age. A person whose diet and exercise don't change with advancing years can develop a weight problem.

A forty-two-year-old woman has taught dancing for the past twenty years. She can expect
(1) never to gain weight if she maintains this program
(2) to increase her weight greatly by age 50
(3) to reduce her caloric intake or put on weight
(4) to increase her caloric intake and remain the same weight
(5) to exercise more

As metabolism slows down with age, even an active person will have to reduce his or her caloric intake in order to maintain the same weight, choice (3). Choices (1), (2), and (4) are incorrect. Choice (5) is not relevant to the question.

The paragraph below gives another example of application of ideas.

EXAMPLE

You may want to be familiar with the two different scales that are used to measure temperature. In the United States, we are most familiar with the Fahrenheit scale (° F), with water freezing at 32° F and boiling at 212° F. Another scale that is coming into more frequent use is the Celsius scale, with water's freezing point at 0° C and its boiling point at 100° C. Use the diagram to answer the following questions.

Paraffin, a waxy solid, begins to melt at 90°F. This is the same as which of the following?

(1) 32°C
(2) 42°C
(3) 46°C
(4) 48°C
(5) 90°C

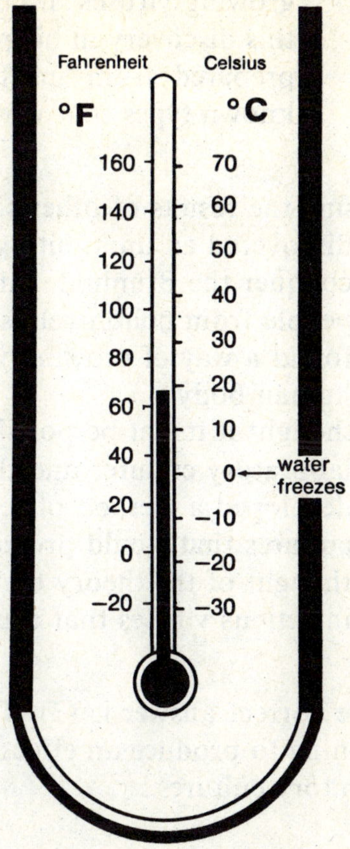

(1) is the correct answer. If you find 90°F on the Fahrenheit side of the thermometer, you will see that 32°C is the closest equivalent.

Inductive Reasoning

Inductive reasoning is used when a person makes generalizations based upon a specific experience and then applies this generalization to other experiences. Since the generalization is made from one example, it leads to a probable conclusion rather than an absolute conclusion. The following reading passage uses inductive reasoning.

EXAMPLE

A drop of water does not seem to be composed of particles. From a distance, the sand that makes up a beach seems to be one large mass. However, up close, we can see that sand consists of small grains. Perhaps, then, water is made up of particles too small to see.

By using inductive reasoning, the author concludes that water, like sand, is made up of tiny particles. Actually, she is referring to atomism, the theory that matter is composed of tiny particles.

Read the following paragraph and answer the question that follows.

EXAMPLE

American virologist Jonas Salk knew from research that there were three types of infectious viruses that cause poliomyelitis: the Brunhilde, the Lansing, and the Leon. He also knew that immunity to one type could not ensure immunity to the others. In an unrelated discovery, Dr. John Enders found a method of growing viruses in cultures outside the human body. Salk used this discovery in his research. Salk was able to develop a vaccine prepared from inactivated poliomyelitis viruses of the three known types that would protect people from the disease.

By using the results of other scientists' research, Dr. Salk
(1) discovered an immunity-giving culture that would conquer the Brunhilde virus and, therefore, protect people from poliomyelitis
(2) found a way of growing viruses in cultures outside the human body
(3) thought it might be possible to grow polio viruses in laboratory cultures and then to develop a polio vaccine
(4) developed a method of researching immunity-giving cultures that would protect individuals from vaccines
(5) thought of the theory that there were three types of infectious viruses that cause poliomyelitis

The correct answer is (3). The passage shows that Salk used inductive reasoning to produce an effective polio vaccine by first growing viruses in laboratory cultures.

Deductive Reasoning

A person who is using deductive reasoning begins with a generalization and draws conclusions about other events using that generalization. Scientists use deductive reasoning to apply general principles to specific situations. Here is an example of deductive reasoning:

Generalization: Scientists have noticed that the increase in oil slicks has had a detrimental effect on birds that depend on the ocean for their food.

Conclusion: If the number of oil slicks continues to increase, the number of sea gulls will decrease.

Now read the following paragraph and answer the question that follows.

EXAMPLE

The elements we find in the earth's crust do not represent the composition of the entire planet. The earth's crust is rich in uranium and thorium, but these two elements are found in greatly reduced amounts just below the earth's surface. If the core of the earth were as rich in uranium and thorium as the crust is, this would cause the earth to melt from the radioactive heat that would be generated.

From this paragraph, one can conclude that
(1) different layers of the earth contain different proportions of elements
(2) there is no uranium or thorium in the earth's crust
(3) only the earth's crust contains uranium and thorium
(4) elements are found in roughly the same proportions throughout the different layers of the planet
(5) the portion of the earth under the surface contains vast deposits of uranium and thorium

Choice (1) is correct. The generalization is stated in the first sentence. You can deduce that, if the composition of the earth's crust does not represent the makeup of the entire planet, then different layers of the earth would have to contain different proportions of elements. The other choices cannot be concluded from the passage.

Now try another example of deductive reasoning.

EXAMPLE

Gravity is the force that draws all objects toward the center of the earth. For example, gravity causes furniture to remain in place in a room. It keeps your feet on the floor. Gravity is the force that must be overcome by airplanes or rockets if they are to leave the surface of the earth.

Using the generalization from this passage, you could conclude that
(1) centrifugal force pulls a space capsule toward the center of the earth
(2) the law of gravity does not apply to rockets or space capsules
(3) rockets and space capsules are subject to centrifugal forces, not gravitational forces
(4) objects above the earth's surface do not experience gravitational forces
(5) the force of gravity pulls space capsules toward the center of the earth

The correct answer is (5). You could conclude from this passage that gravitational forces pull all objects toward the center of the earth. The only response that accurately applies this generalization is answer (5). Choice (4) is incorrect because the forces of gravity are in effect even if an object leaves the surface of the earth.

SCIENCE READING EXERCISE 6

Directions: Read the following passages and answer the questions. Fill in the circle that corresponds to the correct answer.

All car owners are concerned about rust, or iron oxide. Iron oxide is formed when iron and oxygen react with one another. Water and salt speed up this reaction. A protective coating on a metal surface will discourage the formation of iron oxide. The formula for iron oxide is

$$4Fe + 3O_2 \longrightarrow 2Fe_2O_3$$

27. Which of the following can help prevent this reaction from taking place? 27 ① ② ③ ④ ⑤
(1) Lots of water
(2) Exposure to the air
(3) A coat of paint
(4) Salt
(5) Rust can't be prevented.

The survival of living things requires oxygen, water, and moderate temperatures. If the earth were too close to the sun, living things could not survive the heat. If the earth were too far from the sun, they could not survive the cold.

28. Which of the following is a logical conclusion based on this paragraph? 28 ① ② ③ ④ ⑤
(1) Living organisms need only oxygen to survive on earth.
(2) There are only two essential factors that are necessary for survival of life on earth—water and oxygen.
(3) The earth accommodates animals and plants by providing adequate amounts of sunlight, oxygen, and water.
(4) Living organisms on earth can survive without water.
(5) Heat is not a critical factor for life on earth.

29. The main idea of the above passage is that living things require 29 ① ② ③ ④ ⑤
(1) only sunlight and water
(2) atmosphere for animals
(3) carbon dioxide for plants
(4) rainfall for plants
(5) sunlight, oxygen, and water

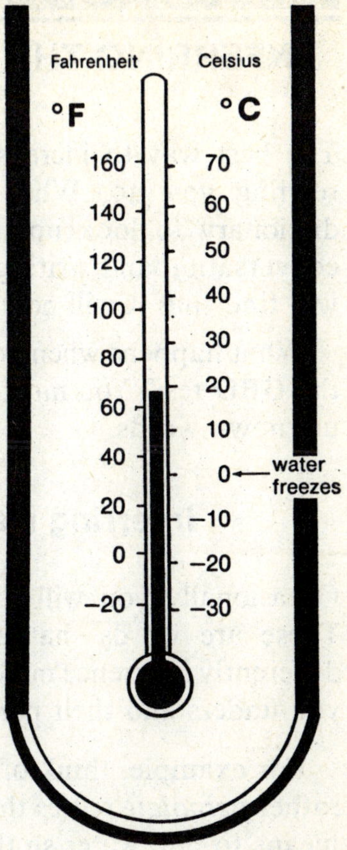

Use the diagram of a thermometer to answer questions 30 and 31.

30. In order to save energy in the winter but still be comfortable, you should set the thermostat in your home at
(1) 4°C
(2) 11°C
(3) 20°C
(4) 25°C
(5) 32°C

30 ① ② ③ ④ ⑤

31. Which of the following is closest to normal body temperature (98.6° F)?
(1) 19°C
(2) 21°C
(3) 32°C
(4) 37°C
(5) 98.6°C

31 ① ② ③ ④ ⑤

For answers and explanations, see page 52.

INFERRING THE MEANINGS OF UNKNOWN WORDS

The best way to increase your vocabulary is to increase the amount of reading you do. When you read, keep a dictionary handy. Use the dictionary to look up unfamiliar words, study them, and use them in conversation and writing. Although this method may seem tiresome, you will find that it will help you increase your vocabulary.

What happens when you don't have a dictionary with you when you take the GED test? You must use other methods to figure out the meanings of unknown words.

Inferring New Meanings of Familiar Words

Occasionally you will find common words used in a scientific context. These are words that are already familiar to you, but they are used differently in science materials. Your understanding of these words will help you understand their particular scientific meanings.

For example, think of the way you use the word *percolate*. Would you rather *percolate* coffee than drink an instant variety? In this case, *percolate* means to boil water so that it will pass through coffee beans. In a science passage, the word may be used in another way.

For example:

EXAMPLE

Magma, or molten rock, percolates beneath the earth, then wells up to burn through solid rock.

Here the word *percolate* is used to describe the formation of a volcano. You may know nothing about the formation of volcanoes, but your familiarity with the word *percolate* will help you understand the process. By applying your knowledge of *percolate*, you can infer the new meaning. Think about the inferred definition of *percolate* as it applies to volcanoes. Magma is so hot it's thought of as boiling up and rising through solid rock.

Using Context Clues to Determine Meaning

To infer the meanings of unknown words, you can frequently use context clues. The context includes the words and thoughts that surround the unknown word. For example, suppose that you do not know the meaning of *resistance*. Which context clues in the following sentences could help you infer the meaning of *resistance*?

EXAMPLE

The filament, a thin wire coil, does not permit electricity to flow through it easily. The electricity must overcome *resistance* in the wire.

The word *resistance* has a special meaning in science material, but you do not need to know the dictionary definition to understand this sentence. The first sentence provides an excellent clue to the meaning. Electricity is not allowed to flow through easily. Something is working against the electricity. There is some kind of opposition. Electricity must overcome opposition. Therefore, in this context, *resistance* means a force that must be overcome.

Valuable clues to the meaning of an unknown word may also exist elsewhere in a paragraph. Read the paragraph below. What clues are there to the meaning of the word *antigens*?

EXAMPLE

The human body has a way of recognizing its enemies. When foreign chemicals enter it, the body attacks the "foe." The most common of the unwanted chemicals come from bacteria, viruses, and other microscopic organisms. The body recognizes these *antigens* and deals with them.

You can infer from the paragraph that *antigens* means
(1) the human body
(2) injections of chemicals
(3) genetics
(4) unwanted substances
(5) desirable bacteria

The correct answer is (4). Although *antigen* is not mentioned until the last sentence, the entire paragraph defines it. The first sentence refers to the body's ability to recognize and attack its enemies. The third sentence talks of "unwanted chemicals." Finally, you were told that the body recognizes the *antigens*. The only choice that defines the *antigen* as an undesirable substance is (4).

As you read the following paragraph, pay close attention to the italicized words. Look for context clues to determine word meanings. Answer the questions that follow the paragraph.

EXAMPLE

All solids do not *liquefy* at the same temperature. If ice remains at a temperature above 32°F for a period of time, it becomes water. Iron ore, on the other hand, becomes *molten* at 2,800°–3,000°F. Wax is solid at room temperature, but *liquifies* at 90°–150°F.

In this passage, *liquefy* means to become
(1) water
(2) different
(3) limp
(4) vapor
(5) liquid

The correct answer is (5). The strongest clue to the meaning of *liquefy* is its similarity to a word you already know—*liquid.* Context clues prove that these two words are related.

According to the passage, *molten* means
(1) melted by heat
(2) decreased in volume
(3) molted by heat
(4) turned to water
(5) heated

The answer is (1). The context clues indicate that *liquefy* and *molten* are related. If iron ore becomes *molten* at 2,800°–3,000°F, a tremendous amount of heat is required. It is reasonable to conclude that *molten* means "melted by heat." Both questions contained choices referring to water. Although water is a liquid, it is not the correct choice. Be sure that your answer is as close as possible to the meaning in the passage.

SCIENCE READING EXERCISE 7

Directions: Read the following passages and answer the questions. Fill in the circle that corresponds to the correct answer.

"Aquaculture" may increase the world's food supply, but there is a danger that it will deplete the ocean's food supply. One popular food is the conch, a shellfish. Scientists are working to replenish the ocean with conch, since it is an excellent source of protein. The conchs are grown in laboratories until they have passed the egg and larva stages. They are kept there until the conchs are large enough to defend themselves against predators.

32. From the context of the passage, you can infer that *deplete* means
(1) eat
(2) pollute
(3) replenish supplies
(4) lower prices
(5) use up

32 ① ② ③ ④ ⑤

33. Since there is a danger of depleting the ocean's supplies, scientists are trying to *replenish*, or
(1) replace them
(2) react to them
(3) rejoin them
(4) repeat them
(5) remind them

33 ① ② ③ ④ ⑤

34. In this passage, *predators* means
(1) sea life that developed before conchs
(2) land animals that prey on conchs
(3) sea life that scientists are trying to preserve
(4) sea life that will replace conchs in our diet
(5) sea life that lives by eating others

34 ① ② ③ ④ ⑤

35. Since *agriculture* means "cultivating or using the soil for food production," you can infer that *aquaculture* means
(1) using space for food production
(2) using the ocean for laboratory experiments
(3) using the ocean for seawater
(4) using the ocean for food production
(5) using laboratories to produce predators

35 ① ② ③ ④ ⑤

Hemophilia is a potentially dangerous disease. A hemophiliac's blood does not coagulate, so even a minor cut can lead to heavy, uncontrolled bleeding. Hemophiliacs must be more cautious in their daily lives than the average person.

36. Based on this passage, we can assume that *coagulate* means
(1) thin
(2) turn red
(3) flow
(4) clot
(5) adapt

36 ① ② ③ ④ ⑤

37. A *hemophiliac* is someone
(1) who gets hurt easily
(2) who easily forms scabs
(3) whose blood does not clot
(4) who never hemorrhages
(5) whose veins are too big

37 ① ② ③ ④ ⑤

For answers and explanations, see page 52.

ANSWERS AND EXPLANATIONS

SCIENCE READING EXERCISE 1

1. (1) The passage discusses the advantages and uses of solar energy. Choices (2), (3), and (4) are supporting details.

2. (4) The passage shows how solar energy could meet our energy needs. Choices (1), (2), and (3) are too general.

3. (5) The passage deals primarily with research activities at Woods Hole. Choices (2) and (3) are too general. Choice (4) is a supporting detail.

4. (1) This statement sums up the main idea of the passage. Choices (3), (4), and (5) are never stated. The passage never implies that there are too many scientific activities, choice (2).

5. (4) The main idea of this passage is that air pollution is a health hazard. The other options support this central theme.

6. (2) While the passage asserts that air pollution is a health hazard, it also suggests that an emphasis on stringent measures to protect the environment could lessen the danger.

SCIENCE READING EXERCISE 2

7. (5) The first sentence suggests there are many causes of dizziness. The paragraph goes on to discuss a variety of these causes. Choices (1) through (4) are all too specific to be correct.

8. (2) Of all the possible causes of dizziness, built-up ear wax or new eyeglasses are the easiest to correct. In addition, they are referred to as less serious than other causes.

9. (4) The passage states that since dioxin bonds itself to soil, the chances that it will seep into water supplies are minimized.

10. (3) "Prior to this [discovery] the EPA had ordered the cleanup of sites containing only 1 part per billion."

SCIENCE READING EXERCISE 3

11. (5) While the article does say that osteoporosis is usually associated with a lack of calcium, it also states that the cause is not known (5). See lines 10–11.

12. (1) One of the effects of osteoporosis is brittle bones, as is mentioned in lines 1–2.

13. (2) While a cure for osteoporosis has yet to be found, it seems that calcium-rich products, choice (2), may reduce the risk of osteoporosis. The other choices are not mentioned as possibilities.

14. (3) Disease-causing organisms and predators limit population growth. You could conclude from the passage that populations that are kept in balance by limiting factors would shrink if the limiting factors increased greatly. The other choices—(1), (2), (4), and (5)—could not be concluded from this passage.

15. (4) Predators limit the growth of the prey animal population. If the predators are removed, then one of the limitations on growth will be removed. Initially, we could reasonably expect an increase in the number of prey animals.

SCIENCE READING EXERCISE 4

16. (4) The last sentence states that, except for birds and mammals, the body temperature of plants and animals is influenced primarily by the external environment. From this statement we can conclude that the body temperature of birds and mammals is influenced by factors other than external environment.

17. (1) Based upon information given in the passage, this is the correct response. Choices (2), (3), and (4) are false. Choice (5) is true, but it is not stated in the passage.

18. (2) The passage states that the body temperature of all organisms, with the exception of birds and mammals, is affected by the external environment. A lizard is the only organism among the choices given that is not a bird or a mammal.

19. (4) The passage states that, in higher-level animals, the embryo grows in the uterus and is nourished by the mother until it is mature enough to be born.

20. (4) The passage discusses the protective function of the embryo's covering. You can conclude that an injury to the protective covering would result in damage to the embryo.

SCIENCE READING EXERCISE 5

21. (4) According to the passage, one of the most important characteristics of aluminum is its resistance to corrosion. You could infer that this makes it an attractive metal to use in the manufacture of cars and airplanes.

22. (2) The passage points out that while other metals are affected adversely by oxygen, aluminum is not. This ability to combine with oxygen to form a tough protective film enables aluminum to resist corrosion.

23. (4) The passage states that the scarred liver cannot absorb and process substances in the blood. Some of those substances would be nutrients. The passage refers to this process as life-sustaining work.

24. (1) Alcohol provides calories but few nutrients. Its lack of nutrition accounts for the malnourished condition of many alcoholics.

25. (1) The passage states that the earth's surface is changing and describes several processes that slowly cause alterations in the earth's crust. Choices (2), (3), and (5) are incorrect. Choice (4) is true only some of the time. The word *always* should make you cautious.

26. (4) Lines 8–12 discuss the effects of the folding of the earth's crust in Scandinavia. Lines 13–15 discuss the effects of volcanoes on the crust in Hawaii and on Mount St. Helens. The passage does not state that these are the only areas affected by crustal strain, choice (1).

SCIENCE READING EXERCISE 6

27. (3) If oxygen can be kept away from iron, then iron oxide can be prevented. Paint can form a protective layer.

28. (3) Only answer (3) is a logical conclusion for this paragraph. You can use inductive reasoning to generalize from the specific items listed in the passage.

29. (5) While all of the choices are true statements about the earth, they serve to support the main idea, choice (5), of this passage.

30. (3) A relatively comfortable thermostat setting that would also conserve heat would be 68°F or 20°C.

31. (4) The equivalent of the normal body temperature of 98.6°F is approximately 37°C.

SCIENCE READING EXERCISE 7

32. (5) The first sentence speaks of the danger of depleting the ocean's food supply. Choices (2) and (5) are dangers. Since the rest of the passage is not about pollution, answer (5) is the better choice.

33. (1) We know that the ocean's supplies are in danger of being used up. If scientists are growing conchs in laboratories, they must be working to replace them.

34. (5) The last sentence states that conchs are kept in laboratories until they are "large enough to defend themselves. . . ." They would need to defend themselves against other sea animals who would eat them.

35. (4) You may know that the word part *aqua* means "water," or you could have inferred its meaning from the passage. Since *agriculture* means "cultivating the soil for food production," *aquaculture* means "using the ocean for food production."

36. (4) Cuts stop bleeding when the blood clots. The passage tells you that, when a hemophiliac gets cut, the bleeding does not stop. You can infer that *coagulate* means "clot."

37. (3) The passage tells you that a hemophiliac is someone whose blood does not coagulate. *Coagulate* means "clot."

Charts, Graphs, and Diagrams

LOOK AT ILLUSTRATIONS CAREFULLY

The main problem with reading scientific charts, graphs, and diagrams is that a person often looks at them with no plan or purpose in mind. As a result, the reader misses some information that is important for interpreting the illustration. When answering questions based on information in a pictorial format, you would be wise to spend a few moments first analyzing the entire illustration.

The title of an illustration gives you the purpose or main idea. Other labeling provides supporting details. You may even be asked to draw conclusions based on a figure.

Here are some specific things that you should do: Read the title, the categories, and dates or amounts. Read all the labels. Then answer the questions in terms of the information given in the figure. If a science passage has an accompanying figure, study the figure first. This will give you insight into the subject before you read the passage.

READING CHARTS

A chart is a figure that organizes information. Charts list, organize, and compare information. Look at the chart below that shows the average weights for men and women.

Figure 1

AVERAGE WEIGHTS FOR MEN AND WOMEN

	Weight in Pounds	
Height	*Women*	*Men*
5'4"	124–136	135–145
5'5"	127–141	137–148
5'6"	130–144	139–151
5'7"	133–147	142–154
5'8"	136–150	145–157
5'9"	139–153	148–160

From the categories at the top of the columns, you can see that information is given on the heights and weights of men and women. If we wanted to find out the average weight of a 5′7″ woman, we would find 5′7″ in the height column and go across that row to the "Women" column. Now try answering the question below.

According to Figure 1 above, what is the average weight of a man who is 5′8″ tall?
(1) 135–145 pounds
(2) 137–148 pounds
(3) 139–151 pounds
(4) 142–154 pounds
(5) 145–157 pounds

To answer this question, find 5′8″ in the height column and go across that row until you find the "Men" column. The correct answer is choice (5), 145–157 pounds.

You must read the headings on charts carefully. Look at the following chart, Figure 2: The Planets.

Figure 2

THE PLANETS

Name	*Distance from Sun in Millions of Miles*	*Time to Orbit the Sun in Years*
Mercury	36	0.24
Venus	67	0.62
Earth	93	1.00
Mars	142	1.88
Jupiter	484	11.86
Saturn	887	29.46
Uranus	1,783	84.00
Neptune	2,794	165.00
Pluto	3,706	248.00

To find the distance from the earth to the sun, find "Earth" in the first column and go across to find the "Distance from Sun." It says "93." We all know that the sun is more than 93 miles away. If you look closely at the heading at the top of the column you can see "in millions of miles," meaning that the earth is really 93 million, or 93,000,000, miles away from the sun.

There may be problems on the GED science test that will require you to compare pieces of data within a chart.

How much farther from the sun is Jupiter than Mars?
(1) 342 miles
(2) 484 miles
(3) 142 million miles
(4) 342 million miles
(5) 484 million miles

Mars is 142 million miles from the sun, and Jupiter is 484 million miles from the sun. To find how much farther Jupiter is from the sun than Mars is, we must find the difference between the two.

484 million miles—Jupiter's distance from the sun
− 142 million miles—Mars's distance from the sun
342 million miles—How much farther from the sun Jupiter is than Mars

The correct answer is choice (4).

CHARTS, GRAPHS, AND DIAGRAMS EXERCISE 1

Directions: Answer the questions below based on the charts you are given. Fill in the circle that corresponds to the correct answer.

Use Figure 2 on the previous page to answer questions 1–3.

1. How many years does it take Uranus to go around the Sun? 1 ① ② ③ ④ ⑤

(1) 84 days
(2) 1,783 years
(3) 29–46 years
(4) 84 years
(5) 8,400 years

2. About how much longer does it take Mars to go around the sun than it takes Venus? 2 ① ② ③ ④ ⑤

(1) .62 years
(2) 1 year
(3) 1.26 years
(4) 1.88 years
(5) 2.50 years

3. The planet that takes approximately twice as long as Uranus to orbit the sun is 3 ① ② ③ ④ ⑤

(1) Pluto
(2) Neptune
(3) Uranus
(4) Saturn
(5) Jupiter

Use Figure 3 to answer questions 4 and 5.

Figure 3

EFFECTIVENESS OF CONTRACEPTION

Percentage of married women who failed to prevent an unwanted pregnancy in the first year of risk			
Douche	38%	Condom	11%
Calendar Method	32%	Monitor Mucus	11%
Foam	22%	Monitor Temperature	10%
Diaphragm	17%	IUD	3%
Early Withdrawal	16%	Pill	1%
Symptothermal	12%	Sterilization	0%

4. Which method of contraception is the least effective? 4 ① ② ③ ④ ⑤
(1) Douche
(2) Early withdrawal
(3) IUD
(4) Pill
(5) Sterilization

5. Short of sterilization, which of the following methods of contraception is the most effective? 5 ① ② ③ ④ ⑤
(1) Douche: 38% failed to prevent pregnancy.
(2) Calendar method: 32% failed to prevent pregnancy.
(3) The pill: only 1% failed to prevent pregnancy.
(4) IUD: only 3% failed to prevent pregnancy.
(5) None of the above

For answers and explanations, see page 73.

READING GRAPHS

A graph is a pictorial way of showing information, using an illustration to show the relationship among facts. Usually a bar or line graph has two sets of facts, one shown down the side (called the **vertical axis**) and the other shown across the bottom (called the **horizontal axis**).

Bar Graphs

The bar graph on the following page shows the different types of emissions (air pollution) by weight. For example, to find how much nitrogen oxide was released into the air, find the bar labeled "nitrogen oxide." Look across the graph from the top of the bar to the label on the vertical axis at the left. Then locate the number on the vertical axis that corresponds to that bar.

Figure 4

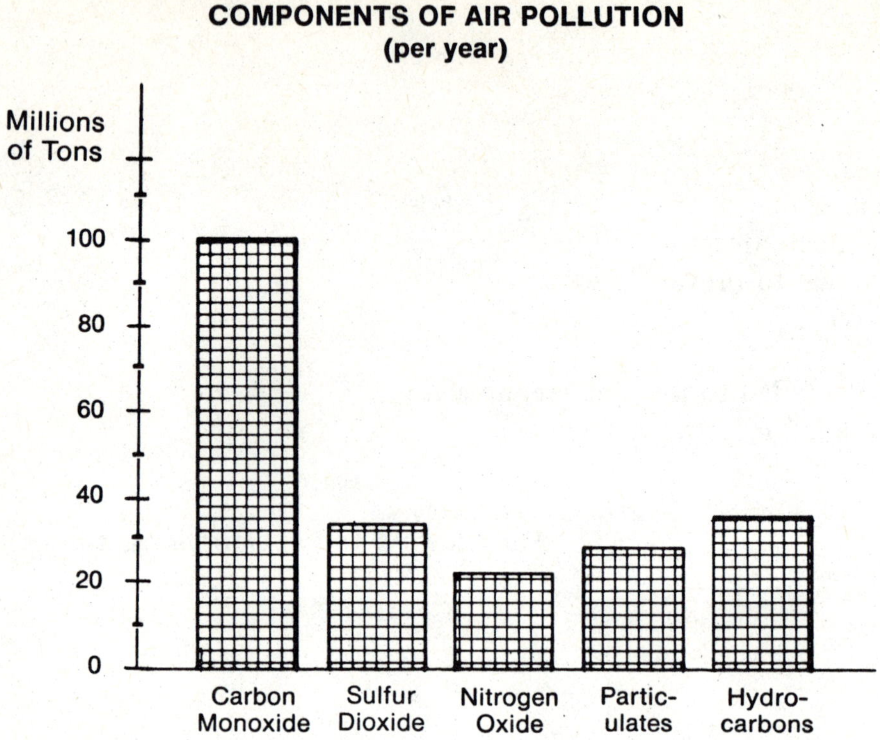

According to the graph, a little more than 20 million tons of nitrogen oxide were released into the air.

A bar graph is useful because it makes it relatively easy to compare data. For example, to determine the most common emission, you would only need to locate the longest bar. The longest bar is the carbon monoxide bar. This bar indicates that 100 million tons of carbon monoxide were released into the environment.

Now try answering another question.

Which of the following substances showed the lowest level of emissions into the air?
(1) Carbon monoxide
(2) Sulfur dioxide
(3) Nitrogen oxide
(4) Particulates
(5) Hydrocarbons

The correct answer is choice (3), nitrogen oxide. The bar for nitrogen oxide is the shortest.

There are also bar graphs that are used to compare different statistics. Figure 5 compares live births with legal abortions. The darker bars represent live births, and the white bars represent legal abortions.

It is very important that you pay attention to the title of the graph. This title has a note that says "rate per 1,000 women." The note means that the 68 you read on the graph does not mean 68 births, but rather 68 live births for every 1,000 women. The 29 legal abortions for 1980 means that for every 1,000 women there were 29 women who had abortions.

Figure 5

ABORTIONS AND LIVE BIRTHS FOR U.S. WOMEN AGES 15-44
(rate per 1,000 women)

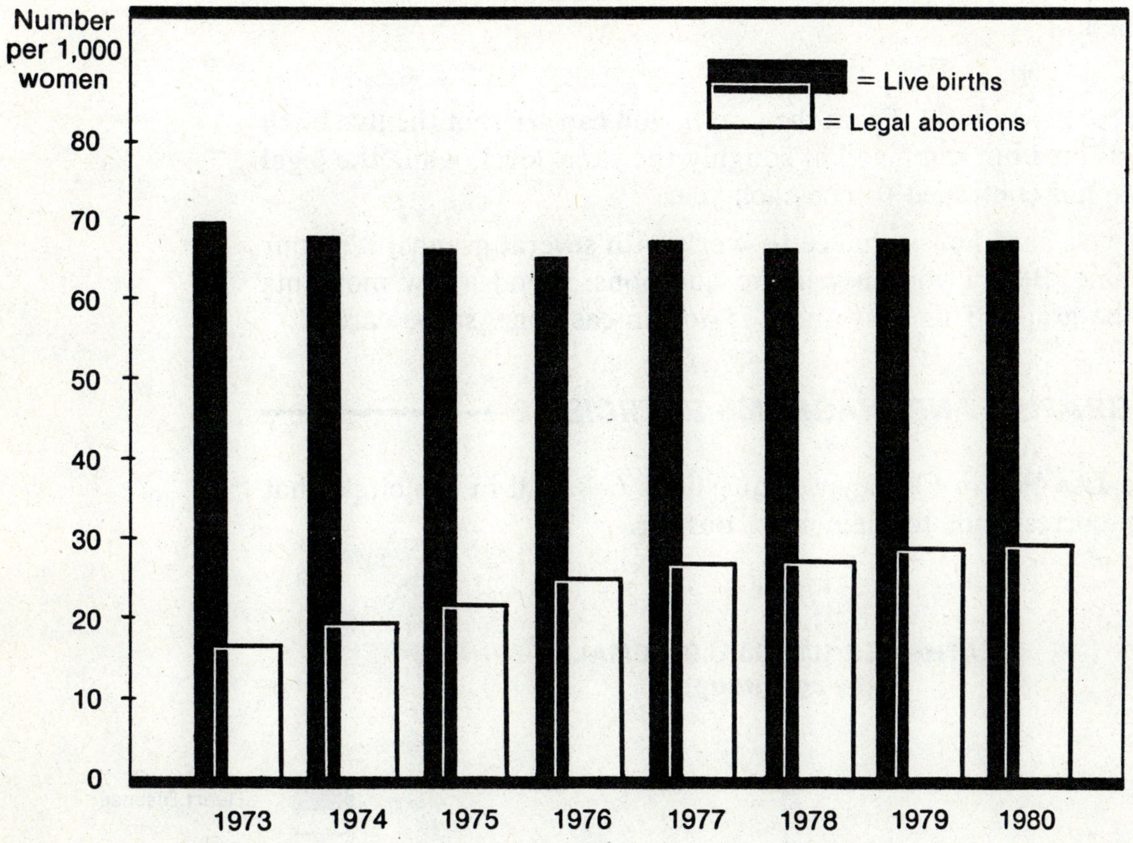

Approximately how many legal abortions were performed in 1976?

(1) 22,000 abortions
(2) 25,000 abortions
(3) 28,000 abortions
(4) 30,000 abortions
(5) None of the above

The correct answer is (5). The graph gives information only on the abortion rate for 1,000 women. It does not tell you how many abortions were actually performed. To find the number of abortions performed, you would need to know the total number of women in the United States. Now try the next question based on the graph.

Which of the following generalizations could be made from this graph?

(1) The rate of live births has remained at roughly the same level while the rate of legal abortions has declined.
(2) The rate of live births has remained at roughly the same level while the rate of legal abortions has increased.
(3) The rates of both live births and legal abortions have remained at approximately the same level.
(4) The rate of live births has increased while the rate of legal abortions has decreased.
(5) The rates of both live births and legal abortions have decreased.

The correct answer is (2). From the graph, you can see that the live birth rate has fluctuated but remained at roughly the same level, while the legal abortion rate has continued to rise each year.

Now that you have had a chance to work with several graphs, it's your turn to try one. Before you answer the questions, spend a few moments examining the graph. This next graph is not an easy one, so be careful.

CHARTS, GRAPHS, AND DIAGRAMS EXERCISE 2

Directions: Use Figure 6 to answer questions 6–8. Fill in the circle that corresponds to the correct answer.

Figure 6

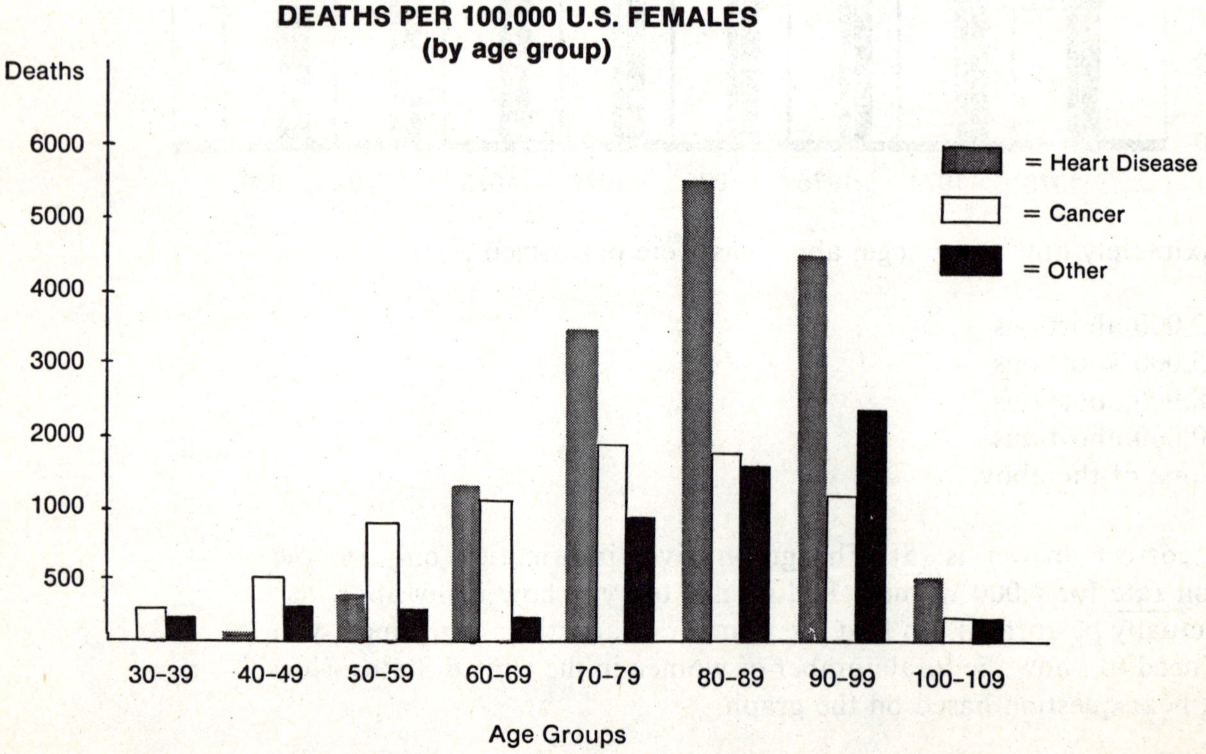

6. Which age group experiences the highest rate of deaths from heart disease?

(1) 60–69 years
(2) 70–79 years
(3) 80–89 years
(4) 90–99 years
(5) 100–109 years

6 ① ② ③ ④ ⑤

7. For which age group below does cancer represent the major cause of death?

(1) 50–59 years
(2) 70–79 years
(3) 80–89 years
(4) 90–99 years
(5) 100–109 years

7 ① ② ③ ④ ⑤

8. Which age group experiences the greatest total death rate?

(1) 50–59
(2) 60–69
(3) 70–79
(4) 80–89
(5) 90–99

8 ① ② ③ ④ ⑤

For answers and explanations, see page 73.

Line Graphs

So far we have worked with bar graphs. Another type of graph that you will encounter on the GED science test is the line graph. Instead of plotting information with bars, a line graph plots data on points that are connected by a line. Line graphs are usually more accurate than bar graphs since they use a continuous curve rather than data chopped up into bars. As on a bar graph, the horizontal and vertical axes of a line graph are used to compare two sets of information.

Figure 7 on the following page graphs the chances, based upon the mother's age, that a mother will have a child with Down's syndrome.

Figure 7

INCIDENCE OF DOWN'S SYNDROME BY MATERNAL AGE

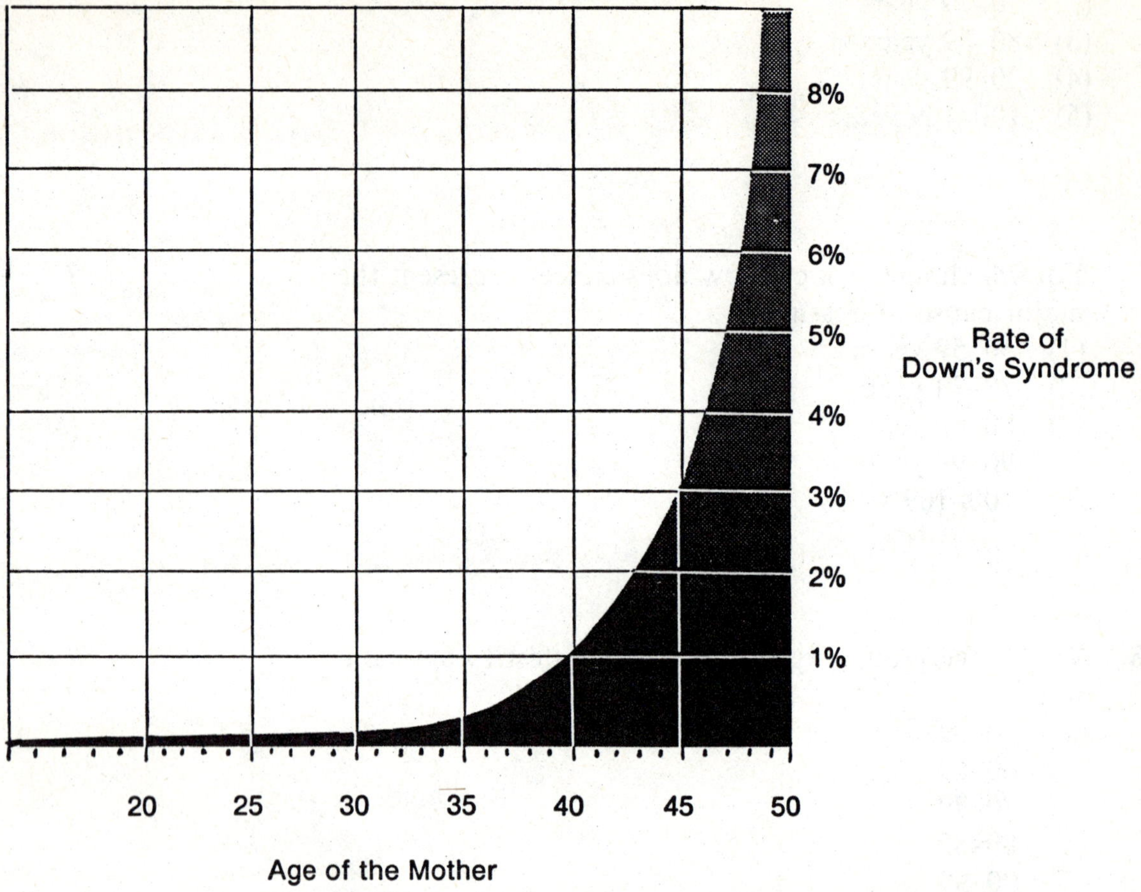

If you wanted to find out at what age a mother would have a 5% risk of having a child with Down's syndrome, you would find 5% on the vertical axis and use your finger or eye to go over to a point on the line that corresponds to 5%. Then, look down to where this corresponds to the point on the horizontal axis. 5% is at 47 years.

Now try the following questions.

At what age does a mother have a 2% risk or 1 in 50 probability of having a child with Down's Syndrome?
(1) 40
(2) 43
(3) 45
(4) 46
(5) 48

The correct answer is (2). A mother at age 43 will have a 2% risk of having a child with Down's syndrome.

At which of the following ages does a mother have the greatest likelihood of having a child with normal intelligence?
(1) 30
(2) 35
(3) 40
(4) 45
(5) 47

The correct answer is (1). For a mother to have the greatest likelihood of giving birth to a child with normal intelligence, she would have the least likelihood of having a child with Down's syndrome. Of the ages listed, 30 is the one that has the smallest probability of Down's syndrome and the greatest probability of having a child with normal intelligence.

There are line graphs that use more than one line, allowing different types of information to be compared. In Figure 8 below, the graph compares the weights of infants. For example, if a nine-month-old infant weighs 22 pounds, she would be considered a large infant. If you follow the line labeled "22" across, you would find that it falls between two lines, in a range labeled "Large."

Figure 8

GROWTH CHART BY WEIGHT

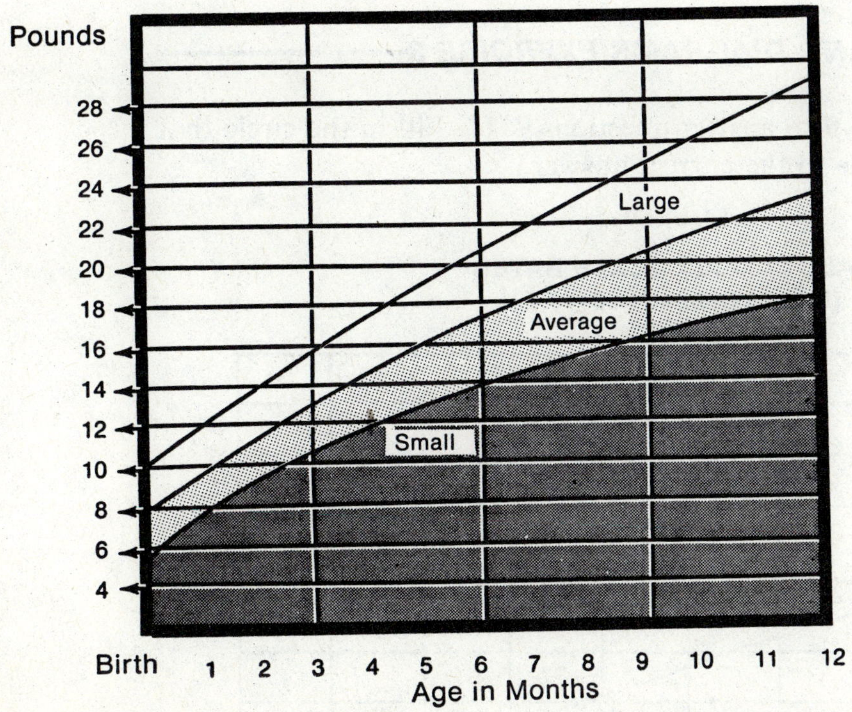

An infant with a birth weight of five pounds would be considered to be
(1) very large
(2) large
(3) average
(4) small
(5) none of the above

The correct answer is (4). An infant with a birth weight of five pounds would be considered small. Find the label "Birth" on the horizontal axis. Look up the vertical axis until you come to the point that stands for five pounds. This point is in the section labeled "Small."

Which of the following weights would be considered average for a six-month-old infant?
(1) 18 pounds
(2) 15 pounds
(3) 12 pounds
(4) 11 pounds
(5) 9 pounds

The correct answer is (2). A six-month-old infant weighing 15 pounds would be considered average in weight.

CHARTS, GRAPHS, AND DIAGRAMS EXERCISE 3

Directions: Use Figure 9 to answer questions 9–11. Fill in the circle that corresponds to the correct answer.

Figure 9

MALE CANCER DEATH RATES
(United States 1930–1975)

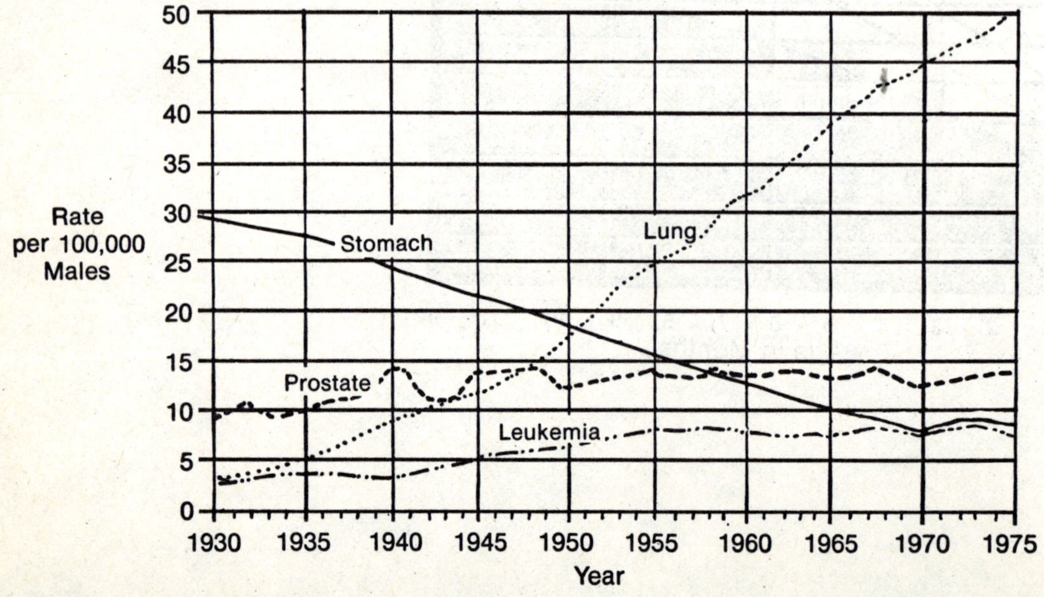

9. For which type of cancer did the death rate drop consistently from 1930 to 1975?

(1) Leukemia
(2) Lung cancer
(3) Stomach cancer
(4) Cancer of the prostate
(5) None of the above

9 ① ② ③ ④ ⑤

10. Which of the following generalizations can be made based on this graph?

(1) The rate of male deaths due to stomach cancer has decreased at the same rate as the other types of cancer.
(2) Stomach cancer is now more common than it was in the 1940s.
(3) Cancer of the prostate will be cured within the next five years.
(4) The rate of leukemia doubled between 1940 and 1970.
(5) The rate of male deaths due to lung cancer has increased dramatically in the past 40 years.

10 ① ② ③ ④ ⑤

11. If the increase or decrease in the rate of deaths from stomach cancer were the same as the increase or decrease in the rate of deaths from lung cancer, which form of cancer would have been the most prevalent in 1975?

(1) Cancer of the prostate
(2) Lung cancer
(3) Leukemia
(4) Stomach cancer
(5) Liver cancer

11 ① ② ③ ④ ⑤

For answers and explanations, see page 73.

READING DIAGRAMS

A diagram is a drawing that shows the parts of an object. Diagrams are also used to show how an object functions. To understand a diagram, it is very important that you carefully read the title and all the labels.

The relationship of an embryo to the uterus is shown in Figure 10 below. The labels indicate several parts of the uterus. Before you try to answer any questions about a diagram, spend a few moments examining the diagram and reading the labels. These pieces of information should enable you to understand the diagram as a whole before you answer the questions. Now try answering the question below.

Figure 10

THE EMBRYO IN EARLY PREGNANCY

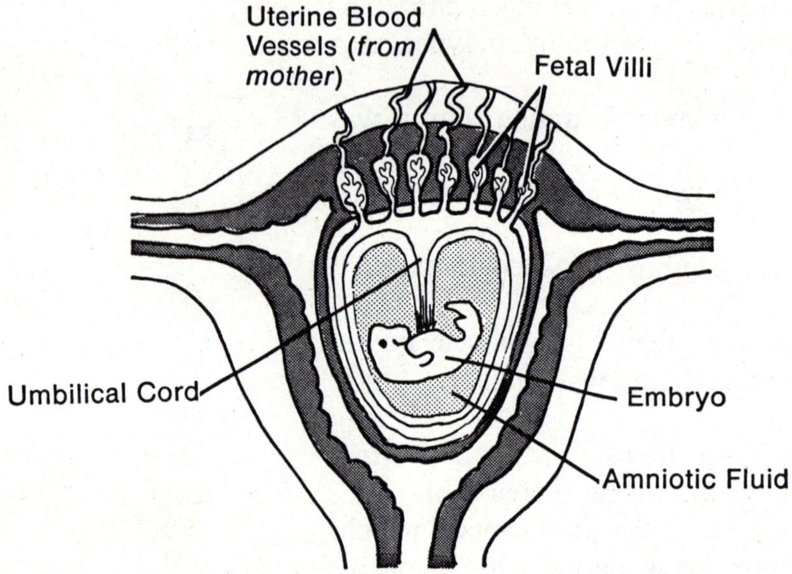

From the diagram, we can conclude that the embryo is surrounded by
(1) amniotic fluid
(2) the umbilical cord
(3) uterine blood vessels
(4) fetal villi
(5) none of the above

The correct answer is (1). The embryo is surrounded by amniotic fluid. After you read the question, you should go back to the diagram. First you should find the embryo in the picture. Then look at the area surrounding the embryo. The line from that area is labeled "amniotic fluid."

Now try answering another question based on Figure 10.

The uterine blood vessels transport blood directly to the
(1) embryo
(2) umbilical cord
(3) fetal villi
(4) uterine wall
(5) amniotic fluid

The correct answer is (3). The diagram shows that the uterine blood vessels are connected directly to the fetal villi.

A diagram of a human heart is shown in Figure 11. As in the examples just discussed, be sure that you carefully examine the diagram and read its labels. Spend a couple of moments studying this diagram so that you will have a good idea of how the heart functions before answering the following questions.

Figure 11

THE HUMAN HEART

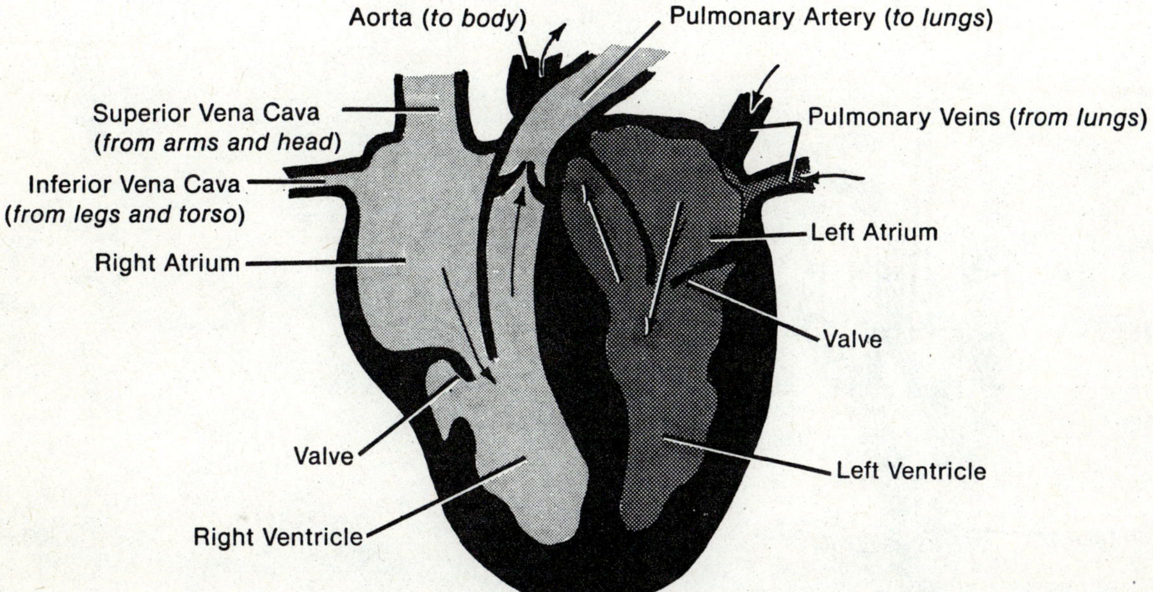

From the right ventricle, the blood is pumped to the
(1) right atrium
(2) lungs
(3) pulmonary veins
(4) left ventricle
(5) legs and torso

The correct answer is (2). From the right ventricle, the blood is pumped through the pulmonary artery to the lungs. The arrow in the area labeled "right ventricle" indicates the flow of blood to the pulmonary artery. This is not given as a choice, but if you look carefully, you will see that the label says "(to lungs)" next to "Pulmonary Artery."

You are on your own for the next two diagrams. With the practice you have had you should be able to do well. Remember to spend a few moments examining the diagram and reading the labels before starting.

CHARTS, GRAPHS, AND DIAGRAMS EXERCISE 4

Directions: Answer the questions below based on the diagrams you are given. Fill in the circle that corresponds to the correct answer.

Use Figure 12 to answer questions 12 and 13.

Figure 12

***FISSIDENS ADIANTOIDES:* MAIN PARTS AND LEAF**

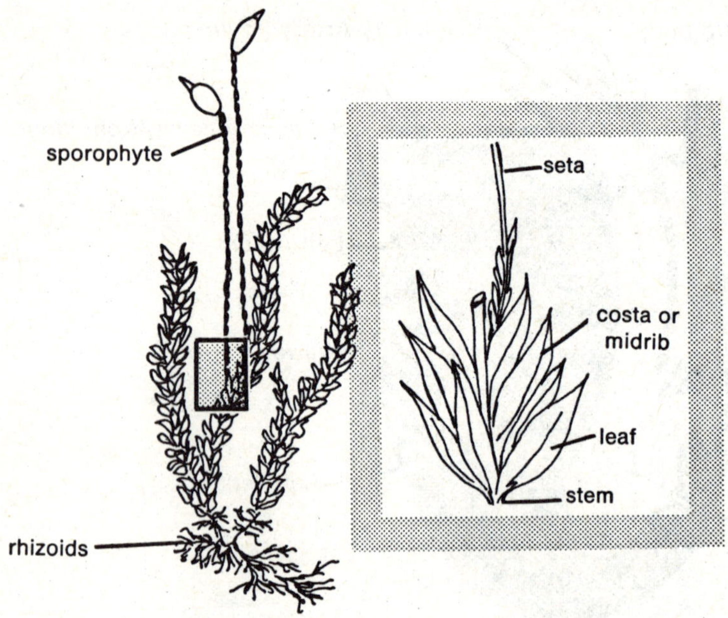

(a) entire plant (b) enlarged portion

12. According to the diagram of *Fissidens adiantoides,* **12** ① ② ③ ④ ⑤
(1) the upper leaves are larger than the lower leaves
(2) the leaves grow out of the stem
(3) the costa is above the midrib
(4) the *Fissidens adiantoides* has only one sporophyte
(5) the seta grows on a separate plant

13. From the diagram, the reader can conclude that **13** ① ② ③ ④ ⑤
(1) the stem of this plant is always erect
(2) the *Fissidens adiantoides* is between 8 and 15 cm tall
(3) another term for the *costa* is *seta*
(4) rhizoids are roots or rootlike parts of the plant
(5) the seta is a bristlelike portion of the plant

Use Figure 13 to answer questions 14–16.

Figure 13

COMMON PUMP

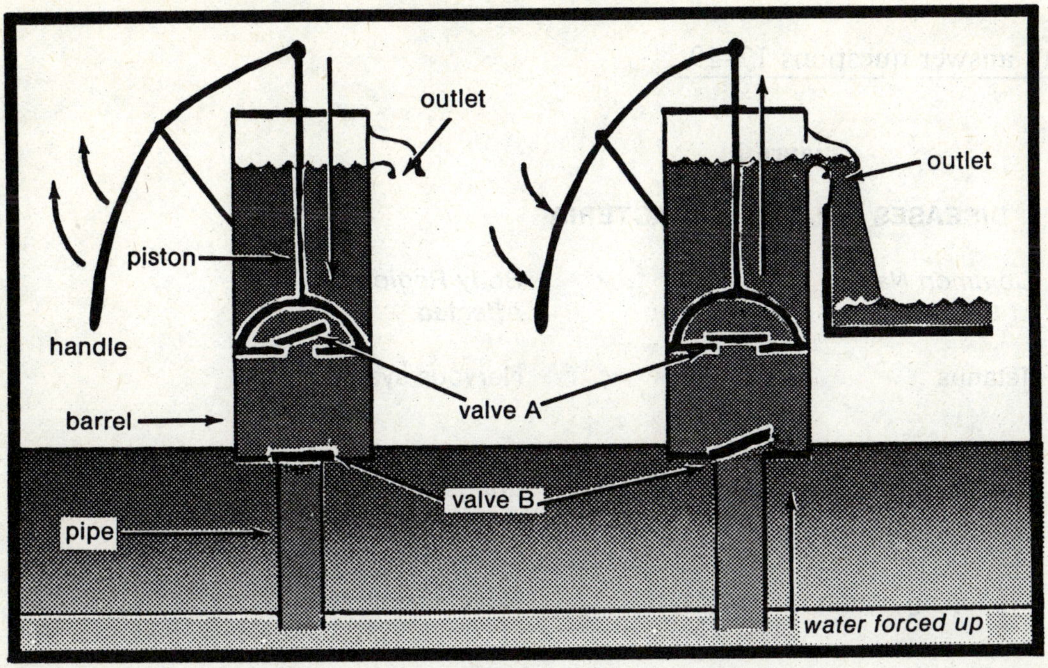

14. When will water be forced up from the source? 14 ① ② ③ ④ ⑤
(1) When valve A is open and the piston descends
(2) When both valves are closed
(3) When both valves are open
(4) When valve A is open and valve B is closed
(5) When valve A is closed and valve B is open

15. When the handle is going up, the 15 ① ② ③ ④ ⑤
(1) piston is also rising
(2) water is entering the barrel
(3) valve A is open
(4) valve B is open
(5) water is being forced up from the source

16. As water is forced out of the outlet, which of the following is also happening? 16 ① ② ③ ④ ⑤
(1) The handle is going up.
(2) Both valves are open.
(3) Water is forced up from the source.
(4) The piston is descending.
(5) Valve A is open.

For answers and explanations, see page 74.

CHARTS, GRAPHS, AND DIAGRAMS REVIEW

Directions: Use the following charts, graphs, and diagrams to answer the questions. Fill in the circle that corresponds to the correct answer.

Use Figure 14 to answer questions 17–19.

Figure 14

DISEASES CAUSED BY BACTERIA

Bacteria	*Common Name of Disease*	*Incubation*	*Body Region Affected*
Clostriclium tetani	Tetanus	3-21 days	Nervous system
Diploccus pneumoniae Mycoplasma pneumoniae Pneumococci	Pneumonia	Varies with type	Lungs
Hemophilus pertussis	Whooping cough	7–14 days	Nose, throat, lungs
Neisseria gonorrhoeae	Gonorrhea	3-9 days	Reproductive organs
Streptococcus	Septic sore throat	3-5 days	Nose and throat
Streptococcus scarlantae	Scarlet fever	2-5 days	Throat
Tubercle bacilli	Tuberculosis (TB)	Few days to years	Lungs and other organs

17. Which disease is caused by the bacteria *Hemophilus pertussis*?
(1) Whooping cough
(2) Scarlet fever
(3) Tetanus
(4) Septic sore throat
(5) Tuberculosis

17 ① ② ③ ④ ⑤

18. Which of the following diseases affects the reproductive organs?
(1) Pneumonia
(2) Scarlet fever
(3) Gonorrhea
(4) Whooping cough
(5) Tetanus

18 ① ② ③ ④ ⑤

19. Which of the following diseases could have a very long incubation period?
(1) Septic sore throat
(2) Scarlet fever
(3) Tetanus
(4) Whooping cough
(5) Tuberculosis

19 ① ② ③ ④ ⑤

Use Figure 15 to answer questions 20 and 21.

Figure 15

THE WATER CYCLE

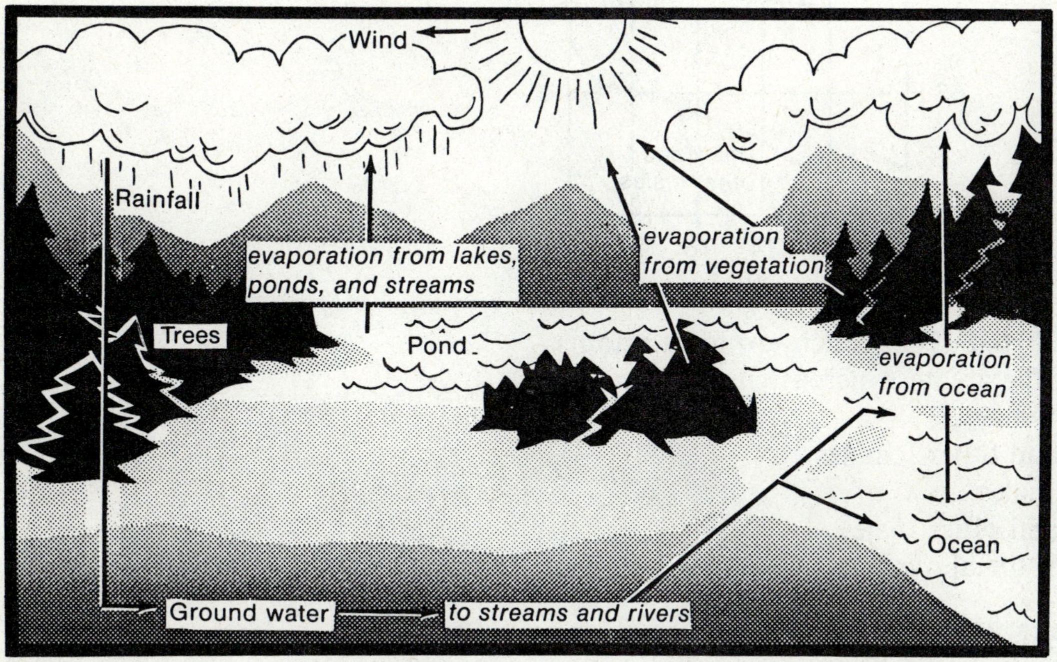

20. After water has entered the ocean, it
(1) falls onto the land as rain or snow
(2) enters the atmosphere through evaporation
(3) reenters the groundwater
(4) enters the ground through evaporation
(5) has finished its usefulness

20 ① ② ③ ④ ⑤

21. Water evaporates from the
(1) clouds and the winds
(2) streams only
(3) rain and snow
(4) trees
(5) vegetation and bodies of water

21 ① ② ③ ④ ⑤

Use Figure 16 to answer questions 22–24.

Figure 16

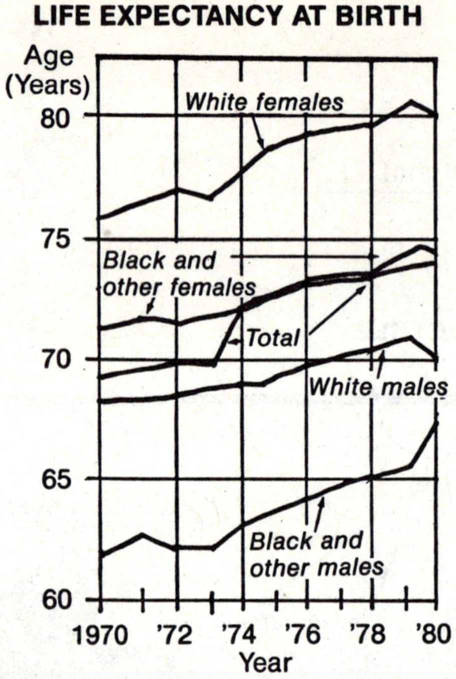

22. Between 1972 and 1980, which group of infants experienced the greatest increase in life expectancy?

(1) White females
(2) Black and other females
(3) White males
(4) Black and other males
(5) None of the above

22 ① ② ③ ④ ⑤

23. A black female born in 1979 would have a life expectancy of about

(1) 76 years
(2) 74 years
(3) 72 years
(4) 71 years
(5) 70 years

23 ① ② ③ ④ ⑤

24. The gap in life expectancy is closing between which two groups?

(1) Black and other males and white males
(2) White females and white males
(3) Black and other females and white females
(4) White males and black and other females
(5) None of the above

24 ① ② ③ ④ ⑤

For answers and explanations, see page 74.

ANSWERS AND EXPLANATIONS

CHARTS, GRAPHS, AND DIAGRAMS EXERCISE 1

1. (4) When you go from "Uranus" to the "Time to Orbit the Sun in Years" column, you will find 84.00 or 84 years.

2. (3) It takes Mars 1.88 years to orbit the sun. It takes Venus .62 years. To find how much longer it takes Mars to go around the sun, you must subtract:

1.88 years—the orbit of Mars
− 0.62 years—the orbit of Venus
1.26 years—difference between the orbits

3. (2) If you take the time it takes Uranus to orbit the sun, 84 years, and multiply it by 2, you have 168 years, which is only slightly more than the time it takes Neptune to orbit the sun.

4. (1) The douche has a 38% failure rate, meaning that 38% of all women who used it failed to prevent an unwanted pregnancy. This is the highest rate shown and therefore is the least effective method.

5. (3) Only 1 percent of the married women who were using the pill failed to prevent an unwanted pregnancy. Therefore, the pill is more effective than any other form of contraception except sterilization.

CHARTS, GRAPHS, AND DIAGRAMS EXERCISE 2

6. (3) The 80–89 age group experiences the highest rate of deaths from heart disease. The longest "heart disease" bar on the graph is in this age group, representing approximately 5,500 deaths per 100,000 females.

7. (1) In the 50–59 age group there are more deaths caused by cancer than by either of the other causes. In the other groups given in the choices, the number of women dying from heart disease or "other" causes is greater than the number of cancer deaths.

8. (4) The 80–89 age group experiences a higher death rate than any other group. You can find this answer by adding the rate of deaths represented by the bars within each age group.

CHARTS, GRAPHS, AND DIAGRAMS EXERCISE 3

9. (3) The death rate of stomach cancer has dropped consistently. The death rate of lung cancer has increased steadily. Deaths from both leukemia and cancer of the prostate have shown fluctuations with a slight increase.

10. (5) There is no question that the rate of male deaths due to lung cancer has increased dramatically. Choices (1), (2), and (4) are false. Choice (3) may be true, but you cannot generalize this from the graph.

11. (4) If the increases in death rates from stomach cancer had been the same as the increases in rate of death from lung cancer, then stomach cancer, which already started at a higher level than lung cancer, would have been the most prevalent cancer-related cause of death in males by 1975.

CHARTS, GRAPHS, AND DIAGRAMS EXERCISE 4

12. (2) None of the other choices are true according to the diagram.

13. (4) The rhizoids are at the base of the plant and resemble the root structure of a plant.

14. (5) A careful examination of the right portion of the diagram shows that water is forced up only when valve B is open and valve A is closed.

15. (3) A careful scrutiny of the diagram will show that when the handle is pushed up, the piston will descend, forcing valve A to open and valve B to close.

16. (3) For water to come out of the outlet, the piston is rising (the handle is going down), valve B is opened, and more water is forced up from the source.

CHARTS, GRAPHS, AND DIAGRAMS REVIEW

17. (1) Look down the column labeled "Bacteria." When you find *Hemophilus pertussis*, look across the row, and you will see that it causes whooping cough.

18. (3) Look under the heading titled "Body Region Affected," and you will see "Reproductive organs." Look across the row, and you will see "Gonorrhea" under "Common Name of Disease."

19. (5) The incubation period for tuberculosis could last years. None of the others lasts more than three weeks according to the chart.

20. (2) The diagram indicates that water enters the ocean through the ground and streams. Arrows directed upward indicate that evaporation occurs in the atmosphere.

21. (5) This is the only answer that describes all of the sources of evaporation listed on the diagram.

22. (4) According to the graph, the greatest increase in life expectancy rates was experienced by black and other males. No other group experienced as great an increase.

23. (2) A black female, born in 1979, would have a life expectancy of about 74 years. While the markings on the graph are not exact, you can see that it is less than 75 years, but not by much. The closest answer is 74 years.

24. (1) Life expectancy for black and other males is rising while it is falling for white males, closing the gap.

Biology

INTRODUCTION TO BIOLOGY

KEY WORDS

organism—any living thing

classification—the organization of plants and animals in groups based on their structure, growth, and function

environment—all conditions that affect the development of an organism

Biology is the science of living things. It is the study of the origin, growth, development, structure, and classification of plants and animals. Biology is concerned with a wide range of topics, from the structure of a single cell to different **organisms** living together. The two broad subdivisions of biology are botany, the study of plants, and zoology, the study of animals.

The science of biology has been developing since humans first took an interest in living things. From looking at cave art we can tell that the study of life dates back to prehistoric times. The development of biological concepts began with the ancient Greeks, who observed and classified animals. During the seventeenth century, the invention of the microscope revolutionized the study of biology, leading to Robert Hooke's famous description of cells as "small boxes."

In the eighteenth century, a method of biological **classification** was developed. During the nineteenth century, the principles of heredity were discovered and the theory of evolution was introduced. The twentieth-century invention of the electron microscope, which uses electrons rather than light waves to magnify images, has opened even greater frontiers for biologists.

The growth of biology has led to the development of many other specialized sciences: biochemistry, the study of the atoms and molecules that form cells; ecology, the study of relationships between organisms and their **environment**; physiology, the study of the functions and vital processes of plants and animals; and genetics, the study of the inherited characteristics of plants and animals.

As the science of biology matures, scientists will learn more about the relationships between different plants and animals and their environment. Their study will help us to learn even more about the world in which we live.

This biology section will focus on three main areas:

1. Human Structure and Functions
2. Cellular and Molecular Biology
3. Evolution and Ecosystems

Directions: Based on what you have just read, fill in the circle that corresponds to the correct answer.

1. The science of biology can best be described as the study of
(1) the inherited characteristics of plants and animals
(2) diseases
(3) the classification of plants and animals
(4) organisms
(5) cell structure

1 ① ② ③ ④ ⑤

2. The difference between botany and zoology is that
(1) botany is concerned with plants while zoology is concerned with animals
(2) the study of botany is more important since it focuses on animals
(3) botany is concerned with animals while zoology is concerned with plants
(4) botany is concerned with living things, while zoology is concerned with the atoms and molecules that form cells
(5) zoology has only recently developed as a science

2 ① ② ③ ④ ⑤

3. We can conclude that most discoveries are not isolated happenings, but rather the result of related developments such as
(1) cave art and the theory of evolution
(2) the microscope and cell theory
(3) the electron microscope and ecology
(4) Robert Hooke and evolution
(5) ecology and heredity

3 ① ② ③ ④ ⑤

For answers and explanations, see page 113.

BIOLOGY EXERCISE 1

Directions: Read the passage below and answer the questions that follow. Fill in the circle that corresponds to the correct answer.

Louis Pasteur and Microbes

Microbes are forms of life that are visible only with a microscope. Some are even too small to be seen with a normal microscope! These organisms are found in soil, plants, animals, streams, oceans, and the air. The discovery of microbes was important to scientists who were searching for the causes of different diseases. In the mid-nineteenth century, the French chemist Louis Pasteur began to explore the idea that some diseases are caused by microbes, especially bacteria, that multiply in the body.

Pasteur discovered how to grow bacteria and found that different species of bacteria cause different diseases. He developed a method for protecting sheep from a disease called anthrax. He vaccinated sheep with a weakened form of the bacillus bacteria that causes anthrax. The vaccination protected the sheep from the poisonous bacteria because their bodies became accustomed to the weaker bacteria and developed antibodies to fight off the stronger form. This began the practice of immunization.

Pasteur was also successful in isolating viruses, another kind of microbe that may cause disease. He applied this knowledge to develop the first effective cure for rabies.

Pasteur also developed a process for preserving perishable foods, such as milk. He realized that these foods spoiled because of the presence of microbes. He developed the pasteurization process, which destroys harmful microbes with controlled heat. This process makes milk and other perishables safe from dangerous microbes that would otherwise grow unchecked.

4. You can infer from the information in the passage that before the microscope was invented 4 ① ② ③ ④ ⑤
(1) it would have been impossible to treat any illness
(2) milk would not have soured because there were no microbes
(3) scientists were not interested in finding the causes of diseases
(4) there was no laboratory work at all
(5) it would have been difficult to discover the relationship between microbes and disease

5. Rabies shots for dogs are a form of which of the following? 5 ① ② ③ ④ ⑤
(1) Experimentation
(2) Sanitation
(3) Immunization
(4) Pasteurization
(5) Sterilization

6. The application of controlled heat to perishables affects the spread of bacteria. This process is called 6 ① ② ③ ④ ⑤
(1) immunization
(2) sanitation
(3) pasteurization
(4) immolation
(5) classification

For answers and explanations, see page 113.

1. HUMAN STRUCTURE AND FUNCTIONS

SYSTEMS OF THE HUMAN BODY

KEY WORDS

system—a group of organs functioning together

organ—a group of tissues working together to perform a complex function

homeostasis—the coordination and balance among systems of the body

Every machine has several **systems** that keep it running. The human body is no different. The major systems of the human body are the digestive, respiratory, nervous, reproductive, circulatory, skeletal, and muscular systems. All of these systems interact with one another. A malfunction in one system of the body affects all of the others.

1. The digestive process breaks down food into a usable form. When the food is chewed, it passes through a series of **organs**. Digestion begins with the esophagus and ends with the large intestine. During its passage through these organs, the food is broken down by different liquids so that it may be absorbed by blood and lymph vessels in the small intestine. From the small intestine, food is carried to all the body cells by the circulatory system.

THE HUMAN DIGESTIVE SYSTEM

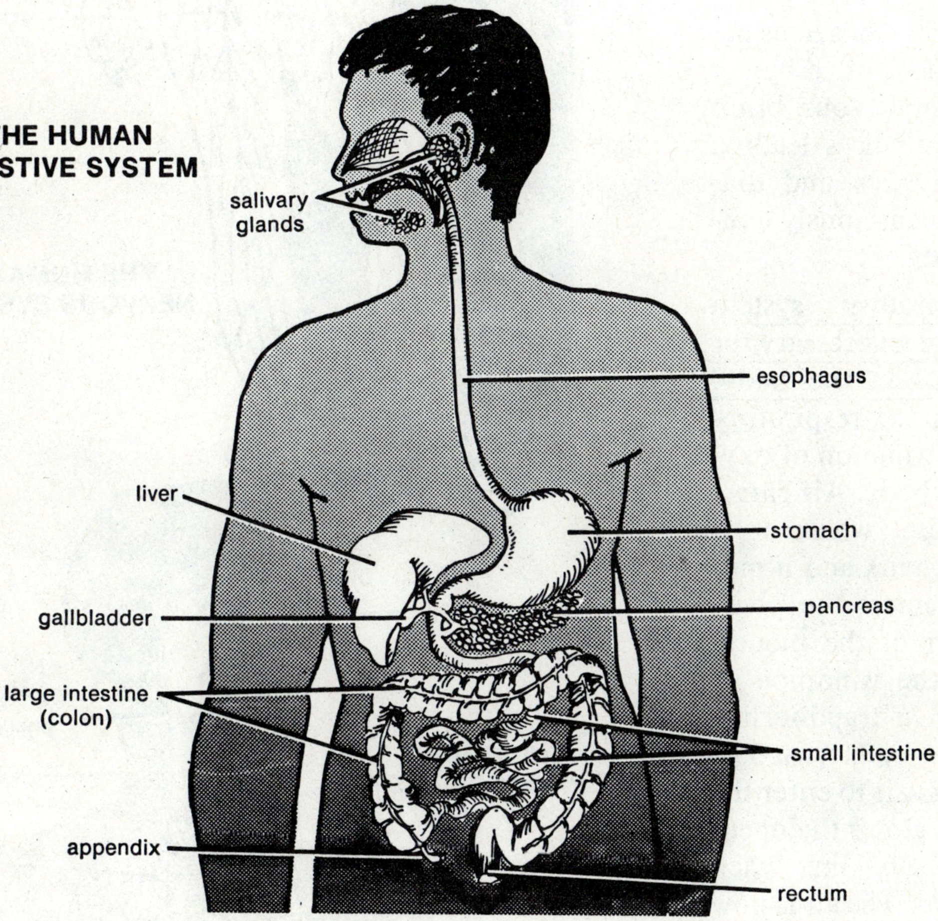

2. The center of the circulatory system is the heart. The heart is a fist-sized muscle divided into two upper and two lower chambers. The upper chambers collect the blood returning to the heart through the veins and pump it to the lungs to absorb oxygen and release carbon dioxide, while the lower chambers pump blood from the lungs into the arteries. The arteries carry blood from the heart to the entire body, using a branching system of progressively smaller arteries, arterioles, and capillaries. The other systems of the body rely upon the circulatory system for nourishment.

3. The nervous system sends messages throughout the body and controls the movement of the muscles. The spinal cord and the brain coordinate the messages carried by the nerves.

When our nerve endings are stimulated, we become aware of how our body responds. Our reflex reaction to pain is to pull back, with nerve impulses traveling to our brain and then back to the muscles.

There are also unconscious activities that are controlled by the brain. For example, our brain controls nerve impulses leading to the chest muscles and diaphragm that unconsciously control our breathing.

4. The respiratory system takes in the oxygen necessary for certain processes in the body. One function of the respiratory system is the distribution of oxygen through the body. Air enters through the nose, where it is cleaned by tiny hairs and a mucous lining. It enters the pharynx, at the rear of the mouth, and goes into the windpipe or trachea through a trapdoorlike muscle called the epiglottis. The epiglottis permits air to enter the windpipe, but it closes the opening to prevent choking when food is swallowed. The air is now in the larynx, or voice box, which is located at the top of the windpipe. The vocal cords (actually two ligaments in the larynx) vibrate when air passes, enabling us to make sounds. The air travels to the lungs by way of the bronchial tubes.

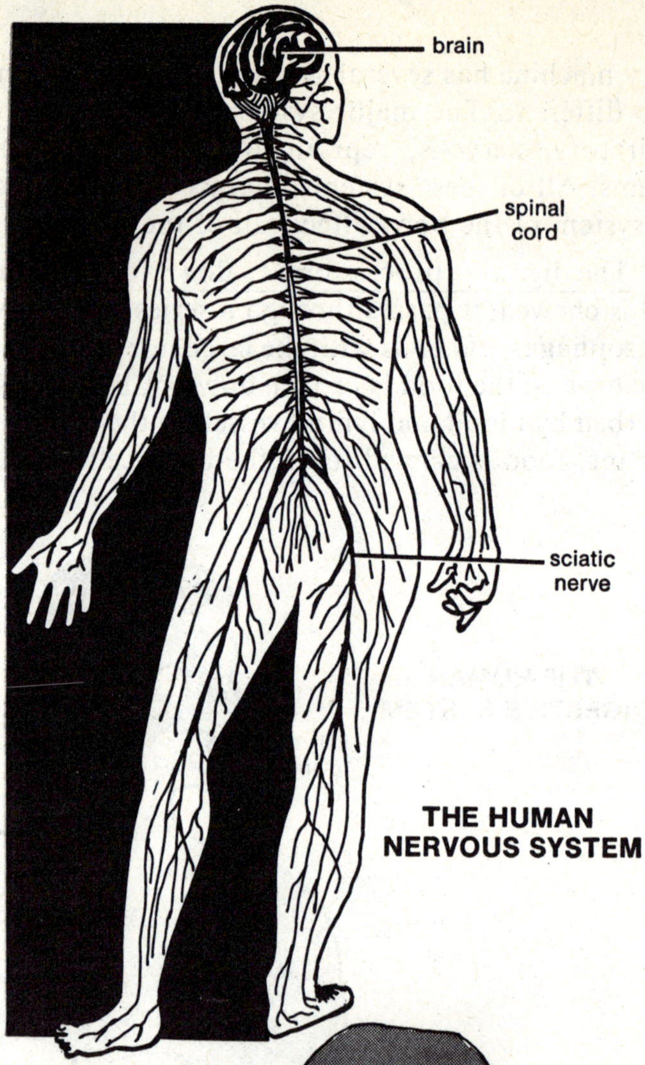

THE HUMAN NERVOUS SYSTEM

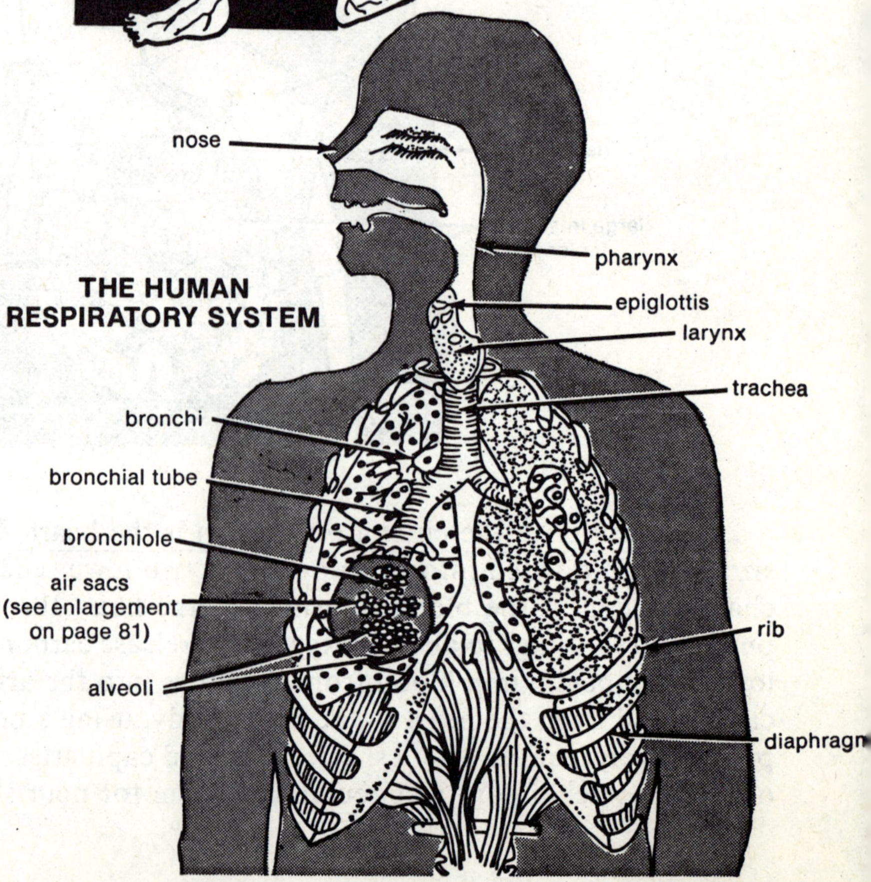

THE HUMAN RESPIRATORY SYSTEM

The bronchial tubes branch into smaller tubes called bronchioles. The air finally flows into millions of air sacs, which are surrounded by a network of capillaries. It is here that blood vessels take the oxygen from the air and send it into the bloodstream. The blood also releases carbon dioxide.

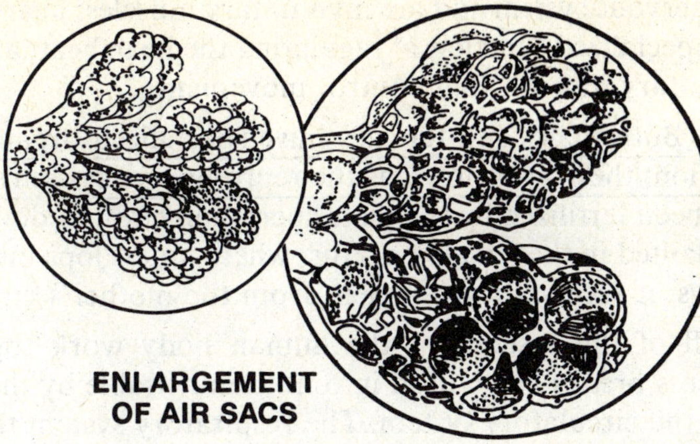

ENLARGEMENT OF AIR SACS

5. The skeletal system is made up of bones, the support structure of the human body. Bones protect vital organs and make movement possible since they provide a frame for the muscles. The skeletal structure, which grows without ever interfering with other working body systems, hardens and becomes brittle with age.

The joints are located where bones come together. There are three principal types of movable joints: ball-and-socket joints, which allow wide circular motion as in the shoulders and hips; hinged joints, which enable back-and-forth motion as in the elbows, knees, jaw, fingers, and toes; and joints that allow limited motion to ankles, wrists, and the spinal column. Ligaments hold bones together.

The muscular and skeletal structures are intricately related. Muscles are described as either voluntary or involuntary. Voluntary muscles are controlled by a conscious decision; involuntary muscles do their work automatically. There are three kinds of muscles: smooth muscles, found in the lungs, intestines, and bladder, which are controlled by

THE HUMAN SKELETON

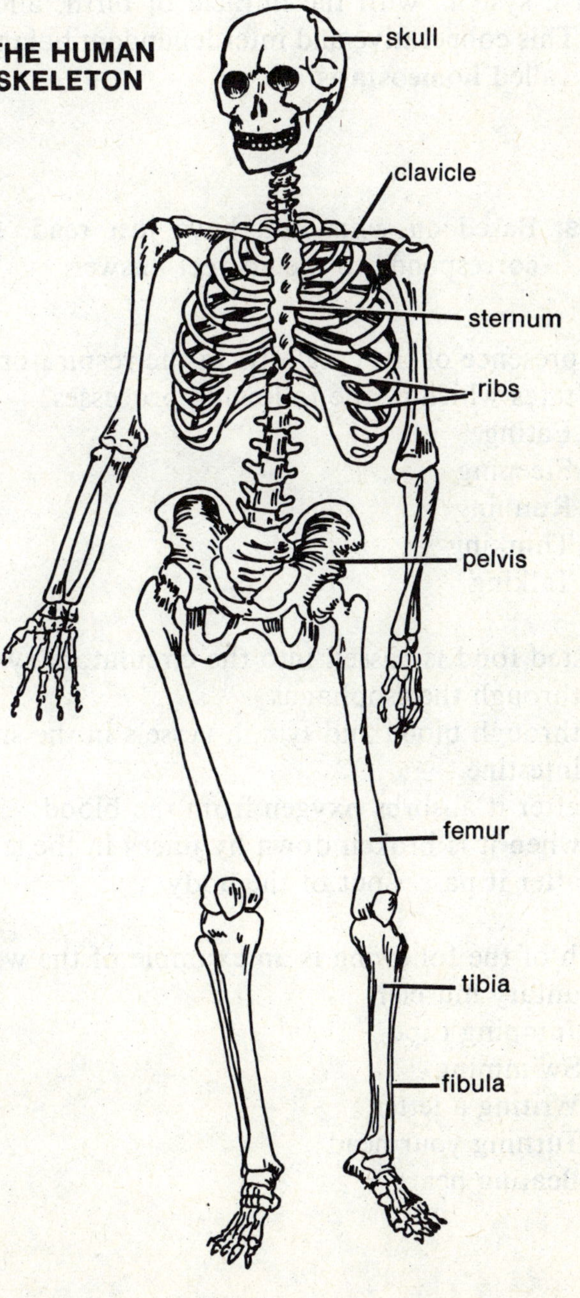

the nervous system and are involuntary muscles; heart muscles, which serve the specialized function of regulating the heartbeat; and striated muscles or voluntary muscles that control movement.

6. Both males and females have specialized organs designed for reproduction; these are known as the reproductive systems. Once the female egg has been fertilized by the male sperm, an embryo develops. This embryo is implanted in the female's uterus, where it develops into a fetus. As the fetus grows, it receives nourishment from the mother's circulatory system.

All of the systems of the human body work together. The digestive process breaks down food into nutrients usable by the body, which are fed into the circulatory system. The respiratory system takes oxygen from the air and passes it into the circulatory system. The circulatory system then supplies nourishment to every cell in the body. The skeletal system provides a structure to protect the organs and make movement possible. The reproductive system, with the miracle of birth, allows humans to create offspring. This cooperative and interdependent balance between the body's systems is called **homeostasis**.

Directions: Based on what you have just read, fill in the circle that corresponds to the correct answer.

7. The presence of the epiglottis in the respiratory tract facilitates which of the following processes?
(1) Eating
(2) Sleeping
(3) Running
(4) Thinking
(5) Talking

7 ① ② ③ ④ ⑤

8. Digested food is passed into the circulatory system
(1) through the esophagus
(2) through blood and lymph vessels in the small intestine
(3) after it absorbs oxygen from the blood
(4) when it is broken down by juices in the stomach
(5) after it passes out of the body

8 ① ② ③ ④ ⑤

9. Which of the following is an example of the work of an involuntary muscle?
(1) Jumping rope
(2) Swimming
(3) Writing a letter
(4) Turning your head
(5) Beating heart

9 ① ② ③ ④ ⑤

10. Breathing through the mouth rather than the nose

(1) isn't possible in humans
(2) causes choking
(3) doesn't properly clean air entering the body
(4) makes much more noise
(5) makes no difference

10 ① ② ③ ④ ⑤

11. Which of the following best describes the relationship between bones and muscles?

(1) They are completely independent of each other.
(2) They are highly interdependent.
(3) Bones do not need muscles for movement.
(4) The bones are more dependent on tendons than on muscles.
(5) Cardiac muscles are the only muscles that are dependent on the skeletal system.

11 ① ② ③ ④ ⑤

12. Which of the following would represent a threat to homeostasis?

(1) A fetus receiving nourishment from the mother's circulatory system
(2) The nervous system sending messages to control muscles
(3) The digestive system breaking down food
(4) Blocking of an artery in the circulatory system
(5) Closing of the epiglottis to prevent choking

12 ① ② ③ ④ ⑤

For answers and explanations, see page 113.

BIOLOGY EXERCISE 2

Directions: Read the following passage and answer the questions. Fill in the circle that corresponds to the correct answer.

Diet and Circulatory Disease

Many Americans have a diet that is rich in sugar and saturated fats. This diet has been linked to a high death rate from heart disease and other circulatory problems. There are several factors that can contribute to circulatory problems: lack of exercise leading to obesity (overweight), heavy cigarette smoking, untreated high blood pressure, and prolonged emotional tensions.

Heart disease is a result of a decrease in the amount of blood supplied to the coronary arteries of the heart muscle. One important factor is believed to be the buildup of fatty deposits—cholesterol—in blood vessels, producing blockages. When the blood flow is completely cut off in a major artery serving the heart muscle, a heart attack occurs. A blockage of an artery that leads to the brain causes a stroke.

High blood pressure, or hypertension, is another heart-related problem. Blood pressure is the force exerted by the blood against the walls of the arteries. Blood pressure usually rises with age because the arteries become less elastic, slowing down the flow of blood. High-fat, high-calorie, and high-salt diets can elevate blood pressure.

Obesity puts an extra burden on the heart and often leads to hypertension. Cigarette smoking can also be a factor since nicotine constricts the blood vessels. If untreated for a period of years, hypertension can affect vital areas of the body, particularly the heart, brain, and kidneys. In many cases, hypertension leads to heart attacks and strokes.

Lack of exercise, which frequently contributes to obesity, and high cholesterol levels can be of critical importance in susceptibility to heart disease. Exercise is believed to promote the growth of collateral or parallel blood vessels that take over for the narrowing main coronary arteries in feeding the heart muscle.

13. From this article, you can conclude that we do have some control over circulatory and heart problems since they are linked to **13** ① ② ③ ④ ⑤
(1) a yearly physical checkup
(2) high blood pressure only
(3) low-fat foods
(4) parallel blood vessels
(5) many factors, including diet, smoking, and lack of exercise

14. Nicotine contributes to hypertension by **14** ① ② ③ ④ ⑤
(1) constricting the blood vessels
(2) coating arteries with plaque
(3) encouraging weight gain
(4) blocking only the collateral blood vessels
(5) building deposits of cholesterol

15. This article states that blood pressure generally rises with age because

(1) tension increases with age
(2) of the cumulative effects of smoking
(3) the body slows down
(4) the arteries become more elastic
(5) the arteries become less elastic

15 ① ② ③ ④ ⑤

For answers and explanations, see page 113.

NUTRITION

KEY WORDS

calorie—a unit of energy; actually the amount of heat needed to raise the temperature of 1 gram of water 1°C

carbohydrates—organic substances composed of sugars and starches; a major source of energy

Calories

We have all heard people talk about counting calories. What they are really talking about is how much energy a particular food can give them. **Calories** are a measure of energy value. Everyone needs calories for energy. We can't live without calories. We need them for everything we do. Our bodies use energy even while we are resting.

Getting energy (calories) from foods is not our only concern. If it were, we could live on candy bars and chocolate cake. We need to choose nutrients that will help our bodies carry out functions of maintenance, growth, repair, and reproduction.

Nutrients

Nutritionists have categorized foods into four food groups to include every nutrient, since a well-balanced diet is essential. The four food groups are meat and poultry, dairy products, fruits and vegetables, and whole grains.

In a well-balanced diet, we should get 45 percent of our daily calories from carbohydrates. **Carbohydrates** are our main source of energy and come in the form of starches or sugars. Bread, rice, potatoes, cereal grains, and bananas are good sources of carbohydrates. Fats represent a concentrated source of food energy and should supply 40 percent of our daily calories. Fats have a high energy content and are easily stored. They can be found in dairy products, meat, and eggs. Protein should supply 12–15 percent of our daily calories. Fish, meat, eggs, and dairy products are common sources of protein.

In addition to calories, our bodies need vitamins, minerals, and water. Minerals are particularly important to the growth and maintenance of the body's systems. Vitamins are necessary for good health.

Nutritionists explain that some foods are more valuable than others. Many foods contain several important nutrients and minerals. As well as providing calories, milk provides protein and fats and the minerals calcium and riboflavin. On the other hand, if you eat a candy bar, you get energy only, since the calories in a candy bar are not accompanied by other nutrients. Nutritionists call them "lone wolf" or "empty" calories.

By choosing our daily menus from the four food groups, we can satisfy our bodies' needs. The actual amount of food we need depends upon an individual's age, sex, build, and level of activity.

Directions: Based on what you have just read, fill in the circle that corresponds to the correct answer.

16. The essay states that not all people need the same number of calories because

(1) the amount needed depends upon a person's level of activity
(2) most people are on diets and need fewer calories
(3) the amount depends upon the foods a person likes
(4) the amount needed depends upon sex, age, build, and level of activity
(5) some people eat more than others

16 ① ② ③ ④ ⑤

17. "Lone wolf" or "empty" calories

(1) are calories that are wasted when you are not eating enough
(2) are calories that are not combined with other nutrients
(3) come from foods that do not have many calories
(4) are calories that include protein, vitamin C, and iron
(5) are usually paired with foods from the milk group

17 ① ② ③ ④ ⑤

THE BASIC FOOD GROUPS

	Meat	*Dairy*	*Fruits and Vegetables*	*Cereal and Bread*
How many servings per day?	2 3 for teenagers	2–3 for children under 9 3–4 for children 9 through teens 3 for pregnant women 4 for nursing mothers 2 for adults	4 (include 1 dark green or deep yellow vegetable for vitamin A, and one citrus fruit or juice for vitamin C)	3 to 4 servings per day (at least one cereal)
What do you get from this group?	Protein Fats Mineral B vitamins	Calcium Phosphorus Protein Vitamin A Riboflavin	Vitamins A and C Minerals Roughage	Starch Bulk Protein Vitamins Minerals

18. According to the chart above, a person who eats no fruits and vegetables has eliminated a major source of 18 ① ② ③ ④ ⑤
(1) fats
(2) carbohydrates
(3) vitamins A and C
(4) calories
(5) protein and fats

19. Which group of people listed below should consume the greatest number of servings from the milk group? 19 ① ② ③ ④ ⑤
(1) Children under nine
(2) Pregnant women
(3) Nursing mothers
(4) Adults
(5) None of the above

20. Which food groups are the best sources of vitamin A? 20 ① ② ③ ④ ⑤
(1) Meat and milk
(2) Meat along with fruits and vegetables
(3) Bread and cereal
(4) Meat and bread and cereal
(5) Fruits and vegetables along with milk

21. Scurvy is caused by a vitamin C deficiency. This weakens the walls of the blood vessels and makes them susceptible to breaking. Which of the following most likely causes scurvy?

(1) Too much meat
(2) Not enough milk
(3) Not enough fruits and vegetables
(4) Too many fruits and vegetables
(5) Too many breads and cereals

21 ① ② ③ ④ ⑤

For answers and explanations, see page 114.

BIOLOGY EXERCISE 3

Directions: Read the passage below and answer the questions that follow. Fill in the circle that corresponds to the correct answer.

Fetal Alcohol Syndrome

Alcohol disturbs an unborn child's development when it is most vulnerable to damage. The damage may be a slight behavioral problem, or it may be the mental retardation or physical deformities that can occur when an infant is born with Fetal Alcohol Syndrome (FAS).

When a pregnant woman drinks, so does her unborn child. The alcohol flows into an unborn baby's bloodstream at a time when the fetus does not have the systems to handle alcohol. The alcohol is poisonous enough to damage the baby's developing organs. Even "social" drinking can be dangerous to the maturing fetus, especially in the first few months of pregnancy.

The damage done to the fetus by FAS is not evident until the baby is born. The FAS baby may show any combination of the following symptoms:

1. low birth weight
2. weak heart or kidney problems
3. learning difficulties
4. failure to develop at a rate comparable to that of other infants

Even when an infant does not have the characteristics of FAS, the pregnant mother's alcohol consumption may become apparent as the child grows older. Researchers call these less apparent consequences "fetal alcohol effects" or "modified FAS." These include:

1. mental retardation
2. abnormally high levels of activity
3. inability to concentrate
4. behavior problems in school
5. abnormal sleep patterns

It has been estimated that FAS is the third most common birth defect. The tragedy is that FAS is easily prevented.

22. The article states that FAS is preventable. What measures are necessary if a pregnant mother wants to prevent FAS in her unborn child? 22 ① ② ③ ④ ⑤

(1) The mother should limit alcohol consumption to occasional "social drinking."
(2) The mother should get lots of sleep.
(3) The mother should limit herself to two ounces of alcohol each day.
(4) The mother should have a balanced diet.
(5) The mother should not consume any alcohol during the pregnancy.

23. Which of the following symptoms of FAS becomes evident only as the child grows older? 23 ① ② ③ ④ ⑤

(1) Mental retardation
(2) Kidney problems
(3) Failure to develop like other infants
(4) Smaller size than other infants
(5) Heart problems

24. If a woman who is a "social" drinker discovers she has been pregnant for two months, she should 24 ① ② ③ ④ ⑤

(1) maintain her consumption patterns since the damage has already been done
(2) abstain from alcohol for the duration of the pregnancy
(3) refrain from alcohol until the second trimester and then resume her drinking
(4) slowly decrease her alcohol intake until she is drinking only one drink at each social occasion
(5) drink only on a full stomach

For answers and explanations, see page 114.

2. CELLULAR AND MOLECULAR BIOLOGY

CELL STRUCTURE AND REPRODUCTION

KEY WORDS

cell—the basic unit of structure of any organism
protoplasm—the living material in the cell
nucleus—the control center in a cell

Cell Structure

Cells are the basic structural units of all living things. They are made up of **protoplasm,** sometimes called the "substance of life," which is the essential living matter of all animal and plant cells. The diagrams below show the differences and similarities between plant and animal cells.

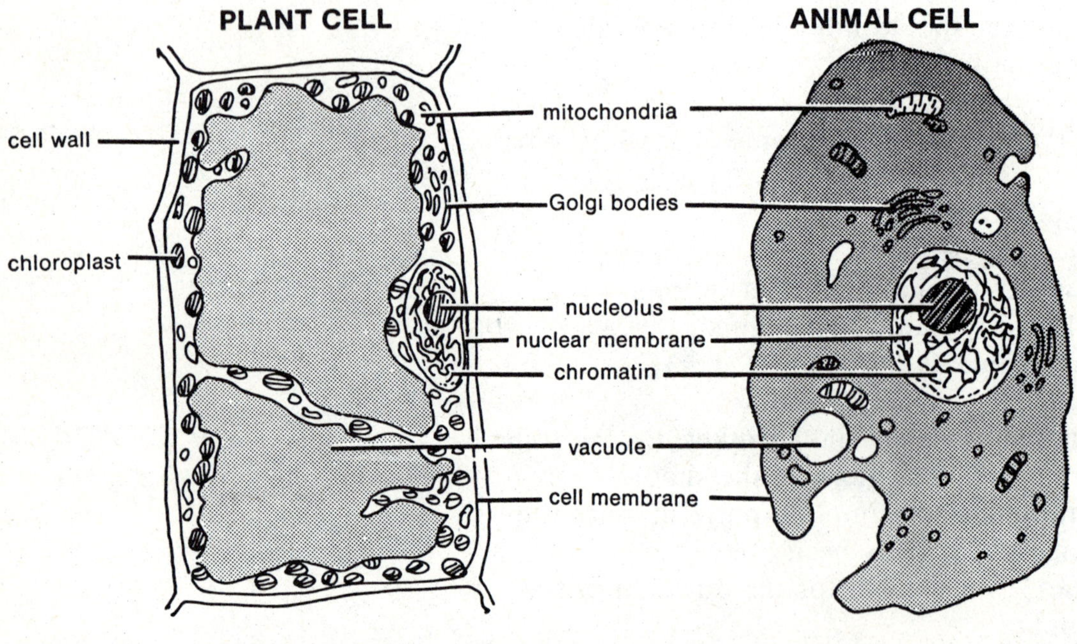

For both types of cells, the cell membrane holds the cell parts together and serves as a screening device for substances passing into and out of the cell. Those substances needed by the cell pass through the membrane into the cell; those substances not needed are not able to pass through the membrane. Waste products are excreted out of the cell through this membrane.

The **nucleus**, a specialized mass of protoplasm, is the central part of each living cell and controls the activities of the cell. A porous membrane surrounds each nucleus, allowing the transfer of material between the nucleus and the surrounding cytoplasm. Within the nucleus, there is a substance called *chromatin*. Globules of this substance are located on superfine threads called *chromosomes*. Chromosomes contain genes, the material that is the basis of heredity.

As you can see from the diagrams, plant cells have several specialized structures not found in animal cells. Chloroplasts, found in the cytoplasm, contain chlorophyll, which is necessary for food production in a plant. A plant produces food through a process called photosynthesis. Plant cells also have a cell wall that surrounds the cell membrane.

Simple and Complex Organisms

The smallest organisms are unicellular, meaning they consist of only a single cell. Bacteria, protozoa, and some algae are unicellular. In a single-cell organism, all functions are carried out by the single cell. In more complex organisms, a division of labor occurs among cells. A group of cells with a common purpose is called a *tissue*. These tissues constitute organs, the specialized bodies you read about in the section on systems of the body. As a plant or animal grows, cells undergo mitosis, or cell division, and assume their specialized duties.

Directions: Based on what you have just read, fill in the circle that corresponds to the correct answer.

25. We can conclude that a major difference between the cells in unicellular and more complex organisms is that cells

(1) in unicellular organisms have cell walls
(2) in unicellular organisms do not contain chloroplasts
(3) in complex organisms carry out all functions
(4) in unicellular organisms have a division of labor
(5) in complex organisms have specialized functions

25 ① ② ③ ④ ⑤

26. A feature of a plant cell not found in an animal cell is

(1) a second nucleus
(2) cell division
(3) protoplasm
(4) chloroplasts
(5) a cell membrane

26 ① ② ③ ④ ⑤

27. The essential living matter of all plant and animal cells is
(1) the nucleus
(2) cytoplasm
(3) chloroplasts
(4) protoplasm
(5) chromatin

27 ① ② ③ ④ ⑤

For answers and explanations, see page 114.

BIOLOGY EXERCISE 4

Directions: Read the passage below and answer the questions that follow. Fill in the circle that corresponds to the correct answer.

The RH Factor

Our genetic inheritance determines not only our blood type but also the presence or absence of the RH factor, an antigen found in the red blood cells of most people. People with the RH factor are known as *RH-positive*, while those who do not have the RH factor are known as *RH-negative*. Approximately 15 percent of all people are RH-negative.

If a person with RH-negative factor has a transfusion of RH-positive blood, an antibody known as anti-RH may build up in the blood plasma. If the person receives a second RH-positive transfusion, the anti-RH will attack the RH-positive red blood cells and cause the blood to clump.

Since the RH factor is inherited, an RH-negative woman must be under careful medical supervision during pregnancy. If the father is RH-positive and the child inherits this factor, some of its blood could mix with the mother's. The number of antibodies built up in the mother could seriously harm the child and would be especially dangerous to a second child.

RH incompatibility in unborn infants is characterized by a destruction of the baby's red blood cells by the mother's antibodies. The baby becomes anemic and jaundiced and, were it not for transfusions, would probably die at or before birth. Upon delivery, the baby's blood is exchanged to get rid of the maternal antibodies. This transfusion is a medical procedure performed to save the baby's life. On rare occasions, blood transfusions have been done before birth.

28. People with the RH factor in their blood cells are said to be
(1) a mutation
(2) an inherited trait
(3) RH-incompatible
(4) RH-positive
(5) RH-negative

28 ① ② ③ ④ ⑤

29. When RH-positive blood is introduced into an RH-negative bloodstream, it can cause the buildup of anti-RH, an agglutinin that reacts with the RH factor. This statement may be considered the reason for
(1) extreme caution during the second pregnancy of an RH-negative mother who has already given birth to an RH-positive child
(2) about 15 percent of all people being RH-negative
(3) tendencies toward leukemia
(4) extreme caution during the second pregnancy of an RH-positive mother who already has an RH-positive child
(5) not allowing RH-positive women to have children

29 ① ② ③ ④ ⑤

30. An expectant mother has RH-negative blood, and her second baby has inherited RH-positive blood from his father. This situation might require
(1) a transfusion of the baby's blood
(2) treatment of the mother for anemia and jaundice
(3) a transfusion of RH-positive blood to the mother
(4) a transfusion of RH-negative blood to the father
(5) a reduction in the number of antibodies in the baby's red blood cells

30 ① ② ③ ④ ⑤

For answers and explanations, see page 114.

GENETICS

KEY WORDS

chromosomes—tiny bodies within the cell nucleus that carry genes

gene—a unit in the chromosomes that transmits hereditary traits

mutation—an inheritable change in the structure of a gene

Genes and Heredity

We have all heard people comment on how children resemble their parents. Comments like "He has his mother's eyes" point out that there are certain characteristics, or traits, that are passed on from parents to their children. We are not surprised that the color of a person's eyes, a person's height, hair color, and other characteristics are passed down through families. How does this happen? The study of heredity, or genetics, has begun to answer this question.

Geneticists, the scientists who study heredity, have found that the cells of living organisms carry **chromosomes** that determine the traits of that organism. Chromosomes are found in the cell nucleus and come in pairs. Human body cells contain 23 pairs of chromosomes.

Genes are tiny particles carried by the hundreds along the length of each chromosome. Genes account for all inherited traits (characteristics). Every person receives two genes for each trait, one from the mother, the other from the father. If those genes are not identical, one gene will be dominant and the other recessive. It is the dominant gene that will appear in any offspring. For example, a blue-eyed mother and a brown-eyed father will usually have brown-eyed children since the gene for brown eyes is dominant.

Cell Division and Reproduction

Scientists believe that each gene is actually a molecule of deoxyribonucleic acid (DNA). This is the chemical substance that makes it possible for an organism to continue to produce another like itself generation after generation.

The only cells in the human body that do not have 23 pairs of chromosomes are the egg cell and the sperm cell. Each of these has only one set of 23 chromosomes, so that when the sperm cell and egg cell unite, the resulting fertilized cell contains 46 chromosomes, or 23 pairs, half from the mother and half from the father.

After the egg cell is fertilized, it begins dividing until the cell division produces the trillions of cells that make up the human body. During each division, the chromosomes in each cell duplicate themselves so that each new cell contains chromosomes that are identical to those of the original cell.

Occasionally a mistake that alters the sequence of the genes will occur in the copying process. A change in a gene, called a **mutation**, produces a changed trait that can be inherited. Many inherited mutations are harmful. Some inherited diseases are muscular dystrophy, a disease characterized by the progressive wasting of the muscles; hemophilia, a condition in which blood fails to clot normally; and sickle-cell anemia.

Directions: Based on what you have just read, fill in the circle that corresponds to the correct answer.

31. Except for sperm and egg cells, every new human cell
(1) has chromosomes identical to the ones in the original cell
(2) is half the size of the original cell
(3) has a different function from the original cell
(4) is entirely different from the cell from which it originated
(5) contains 23 chromosomes

31 ① ② ③ ④ ⑤

32. Chromosomes are found in the
(1) genes
(2) cell nucleus
(3) egg and sperm cells only
(4) original cells only
(5) tiny particles of genes

32 ① ② ③ ④ ⑤

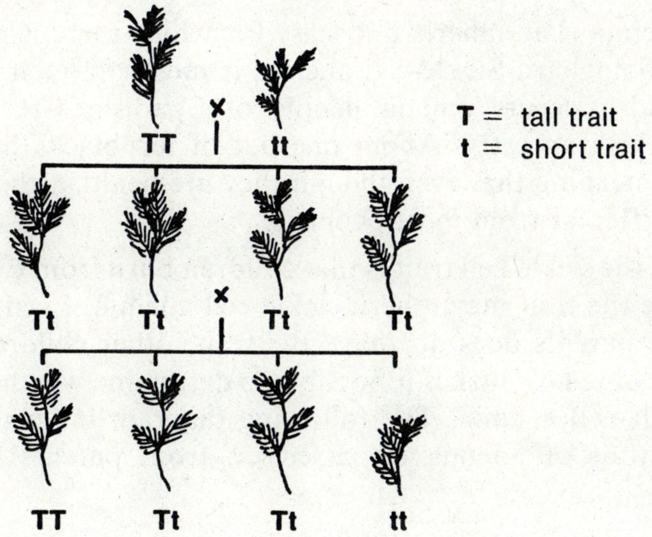

33. The diagram above shows a cross between a tall and a dwarf plant and their first- and second-generation offspring. From this diagram, you could conclude that the trait for short size (t) is
(1) dominant
(2) recessive
(3) a mutation
(4) not found on the gene
(5) harmful to the plant

33 ① ② ③ ④ ⑤

For answers and explanations, see page 115.

BIOLOGY EXERCISE 5

Directions: Read the passage below and answer the questions that follow. Fill in the circle that corresponds to the correct answer.

Sickle-Cell Anemia

A sickle cell is a diseased red blood cell. A healthy red blood cell is shaped like a doughnut, but sickle cells are shaped like a sickle. The hemoglobin in sickle cells releases too much oxygen, and the cells collapse into the shape of a sickle and die quickly, leaving the body without enough red blood cells to meet the demand for oxygen. Sickle cells tend to clump together, clogging the blood vessels and cutting off oxygen to body tissues, especially in the hands, feet, and bones.

Sickle-cell anemia affects the blood cells of about 50,000 Americans. Some of the common symptoms are slow growth, poor circulation, sores that won't heal, swelling of the feet and hands, aching joints, and severe pain caused by clogged blood vessels.

Sickle-cell anemia is an inherited disease for which there is no cure. It is not contagious. Sickle-cell anemia is most common in blacks, but it also occurs among people of Spanish, Greek, Italian, and Turkish ancestry. About one out of ten blacks has sickle-cell trait, meaning that even though they are healthy, their hemoglobin is different from most people's.

The danger of the sickle-cell trait is that children born from two people who have the trait may inherit sickle-cell anemia. Even if one child of the parents does not have the trait, other children may. Prenatal testing now makes it possible to determine whether a fetus has sickle-cell anemia. The following diagram indicates how sickle-cell trait or anemia is passed on from parents to children.

34. The hemoglobin in sickle cells
(1) is highly contagious
(2) is healthy
(3) is not inherited
(4) is shaped like a doughnut
(5) releases too much oxygen

34 ① ② ③ ④ ⑤

HOW PARENTS PASS ON SICKLE-CELL ANEMIA

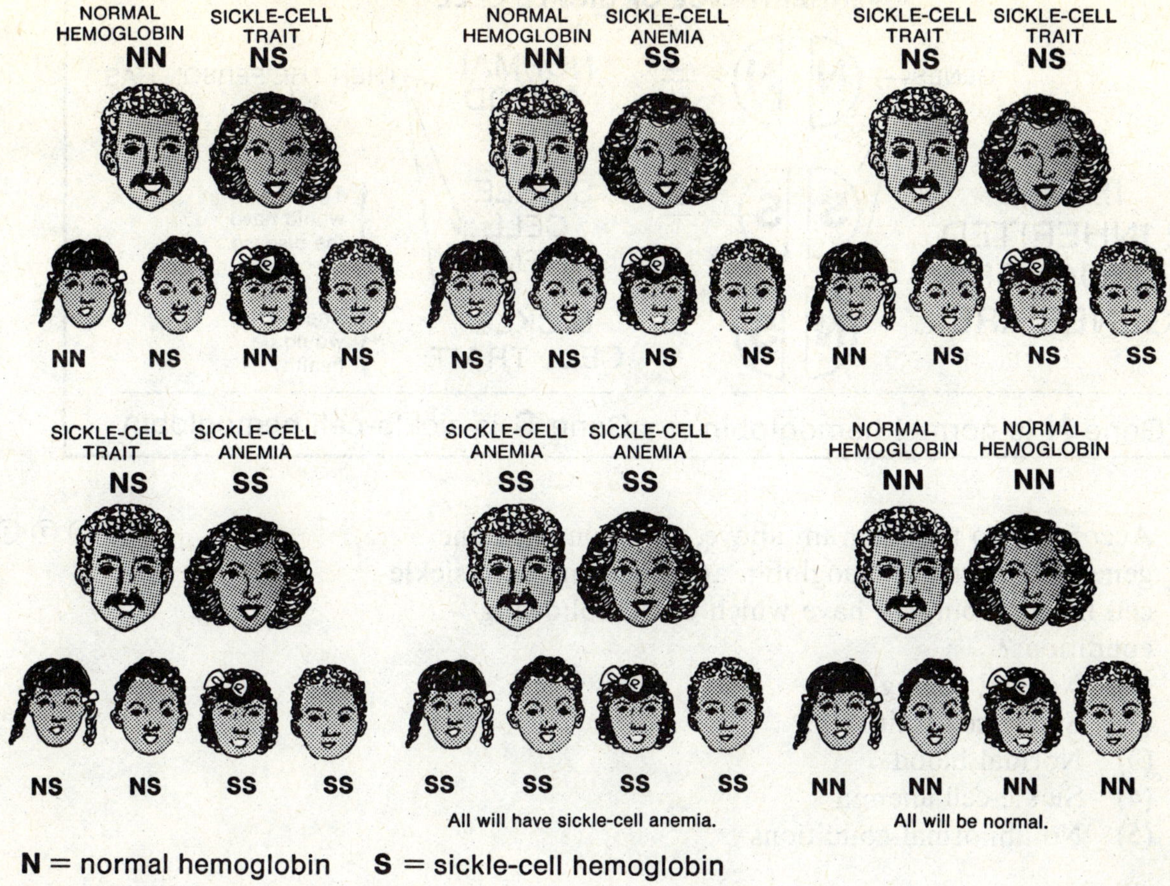

35. According to the diagram above, if one parent has sickle-cell trait and the other parent has sickle-cell anemia, what are the possibilities that they will have a child with sickle-cell anemia? 35 ① ② ③ ④ ⑤

(1) No possibility of sickle-cell anemia
(2) One out of four
(3) Two out of four
(4) Three out of four
(5) The child will definitely have sickle-cell anemia.

36. If one parent has normal hemoglobin and the other has sickle-cell anemia, which of the following conditions is most likely? 36 ① ② ③ ④ ⑤

(1) There is a one in four possibility that any child will have sickle-cell trait.
(2) There is a one in four possibility that any child will have sickle-cell anemia.
(3) There is a 50-50 probability that each child will have sickle-cell trait.
(4) All of the children will have sickle-cell trait.
(5) All of the children will have sickle-cell anemia.

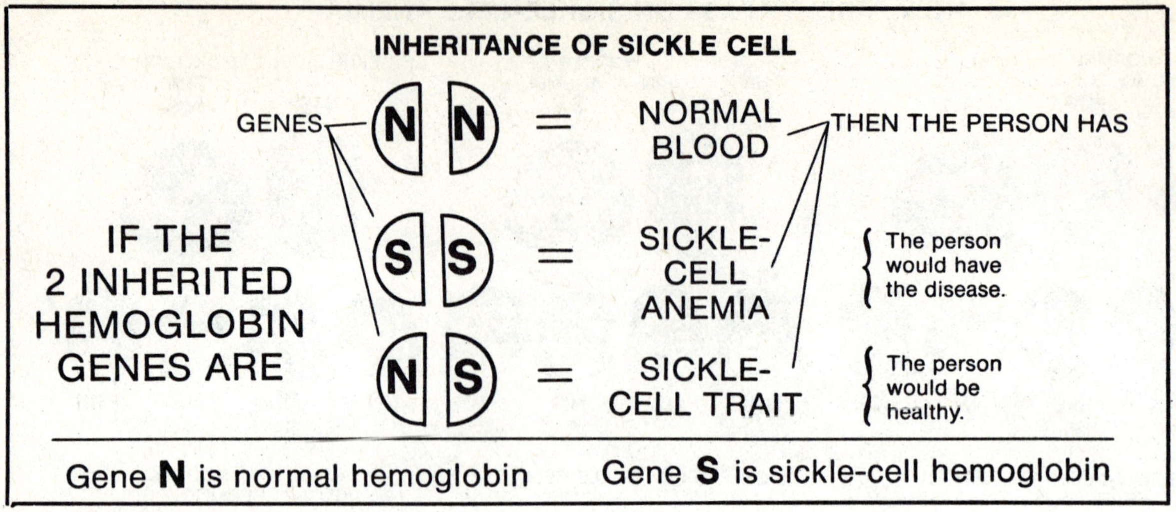

37. According to the diagram above, a person with one gene with normal hemoglobin and one gene with sickle-cell hemoglobin will have which of the following conditions?

(1) Normal hemoglobin
(2) Sickle-cell trait
(3) Normal blood
(4) Sickle-cell anemia
(5) No abnormal conditions

37 ① ② ③ ④ ⑤

For answers and explanations, see page 115.

MICROBES AND MICROBIOLOGY

KEY WORDS

microorganisms—organisms too small to be seen without a microscope

A **microorganism** is an organism that is too small to see with the unaided eye. Most microbes are less than $^4/_{1000}$ inch (0.1 millimeter) across so that the only way to see them is with a microscope. Microbiology is the study of microorganisms. It involves the identification of microorganisms and studies their structure and functions.

Although microorganisms are very abundant, they usually go unnoticed. They do, however, provide much evidence of their presence. Microorganisms cause decay, generate disease, ferment alcohol, raise bread, and can be used to create dairy products. While some microorganisms cause disease, many are helpful to humans.

Among the microorganisms biologists study are (1) protozoa, single-celled animals; (2) algae, primitive plants usually found in water or damp places; (3) fungi, nongreen plants; (4) bacteria, single-celled organisms that can cause serious disease; (5) viruses, which live in a cell of another living thing and cause such diseases as measles, influenza, and common colds; and (6) rickettsias, which cause infectious diseases that are spread by insects, such as Rocky Mountain spotted fever.

Directions: Based on what you have just read, fill in the circle that corresponds to the correct answer.

38. The word *microorganism* means
(1) magnified
(2) organism
(3) biology
(4) tiny living thing
(5) bacteria

38 ① ② ③ ④ ⑤

39. Which of the following generalizations can you make about microorganisms?
(1) They are mainly harmful because they spread diseases.
(2) They are useful mainly to scientific research.
(3) Modern medicine has made them obsolete.
(4) They have many uses in nature and manufacturing because they contribute to decay.
(5) Most of them are not useful to humans because they contain chlorophyll.

39 ① ② ③ ④ ⑤

For answers and explanations, see page 115.

BIOLOGY EXERCISE 6

Directions: Read the following passage and answer the questions. Fill in the circle that corresponds to the correct answer.

Sniffing Out a New Flu Vaccine

A newly developed nose drop vaccine against the most deadly form of influenza virus holds promise as a better method of stopping a flu epidemic before it starts. One or two drops of the new vaccine, squirted into the nostrils of healthy college-age volunteers who were later exposed to the flu, protected all students tested from the muscle aches, coughing, and high fever of flu. The treatment also kept those who contracted a mild, symptomless infection from spreading the virus to others. The findings are especially important to the elderly and those with chronic lung disease, for whom a bout of influenza Type A infection can escalate into fatal pneumonia.

Unlike conventional flu vaccines that rely on injections of a killed virus to trigger an immunity buildup, the new vaccine contains a live virus rendered harmless in the laboratory. Live virus vaccines generally provide longer lasting and more potent protection than killed or inactivated virus vaccines but carry a small risk that they themselves will revert to a virile form and cause disease. The modified virus in the nose drop seems very stable and not prone to such a reversion. Provided the experimental product passes further tests of safety and effectiveness in healthy elderly persons, a commercial vaccine could be ready in three to five years if a company is interested in producing it.

Flu viruses seem to grow best in cells that line the nose, throat, and lungs, and the nose drop vaccine prompts a localized rise in immunity directly at the site of infection. In contrast, conventional injected shots prompt a generalized rise in immune cells and their chemical weapons that circulate throughout the blood, raising respiratory immunity only slightly. Scientists suggest that this difference in the site of action might account for the greater resistance to flu infection found in volunteers given the experimental vaccine.

40. How is influenza spread? — 40 ① ② ③ ④ ⑤

(1) Vaccine
(2) Bacteria
(3) College students
(4) Virus
(5) Muscle aches

41. Why is flu vaccine especially important to the elderly? — 41 ① ② ③ ④ ⑤

(1) The elderly usually have antibodies for the flu.
(2) They are the least susceptible group of people.
(3) The elderly already have strong immunities to the flu.
(4) The elderly have already used up all their immunities.
(5) They risk contracting fatal pneumonia whenever they have the flu.

42. What is the danger of using a live virus vaccine?
(1) The virus may produce an antibody.
(2) It works for only a short period of time.
(3) Live virus vaccines are too strong to use on humans.
(4) It seems to be an ineffective inoculation.
(5) It may cause a disease.

42 ① ② ③ ④ ⑤

43. Flu viruses grow best in the
(1) cells lining the respiratory system
(2) chest and shoulder muscles
(3) cardiovascular system
(4) body's immune system
(5) stomach and small intestine

43 ① ② ③ ④ ⑤

For answers and explanations, see page 115.

3. EVOLUTION AND ECOSYSTEMS

THE THEORY OF EVOLUTION

KEY WORDS

theory—a coherent explanation of a group of experiences or facts

species—a group of living organisms capable of interbreeding

natural selection—the theory of the "survival of the fittest"

Evolution, the idea that more complex forms of life have evolved from simpler forms, has been widely discussed as an important scientific **theory** since the publication of Charles Darwin's *The Origin of the Species* in 1859. However, Darwin was not the first naturalist to consider the idea of evolution. Earlier scientists had noted similarities among **species**, but their work had not produced a well-organized body of evidence, and they had not developed an explanation of how evolution had occurred. Darwin did both.

Darwin's Theory

Darwin's theory consisted of three major ideas. First, he proposed that all living things, some two million different species, developed from a small number of primitive, simple organisms. According to his theory, the gradual changes in these organisms occurred over six hundred million years. As evolution proceeded, some offspring were different from their parents. These offspring also evolved so that succeeding generations became less and less alike. Eventually, the diverse offspring were no longer members of the same species, so they could not interbreed. Through this concept of evolution, Darwin explained how certain species of amphibians developed into reptiles and how certain species of reptiles developed into birds and mammals.

Second, Darwin proposed that transmutation, the orderly change of species, occurs because of **natural selection**. He had read studies that said that populations multiply faster than the available food supply. As a result, there is competition for food. It became clear to Darwin that nature acts as a selective force, allowing the weak to be killed and the strong to survive. Those surviving might further evolve into new species even better suited to a changing environment. The elimination of the less fit, or the "survival of the fittest," is known as natural selection.

Finally, Darwin proposed that organisms that survive can adapt to different conditions. Successful adaptation increases the ability of a plant or animal to survive and reproduce.

Darwin's Insights

Darwin observed adaptation on the Galapagos Islands. There, different species of finches (small seed-eating birds) thrived on different, isolated islands. All the finches had certain characteristics in common, but each species had adapted in a way that would allow it to cope with its surroundings. The main difference among the species was the size and shape of their beaks. Each was suited to the kind of food available on its island. Darwin maintained that those birds whose beaks had not adapted died out.

Structural similarities in different animals indicate that these animals may share common ancestors. For example, the outward appearance of a person's arm, a bat's wing, and a whale's flipper show little resemblance. However, a close look at their skeletons show that they have similar bone structures. Over a long period of time, each developed in a different direction—for life on land, flying, or swimming. The similarities suggest a common ancestor in the distant past.

Darwin's theories were incredible for his time. Prior to Darwin, species had been considered to be unchanged since the beginning of time. Darwin's work, and the work of scientists who followed, attempted to demonstrate that animals and plants have been evolving since the beginning of life.

Scientists are still asking questions about the nature of evolutionary changes and their causes. For more than a century, evolution has been tested against a growing accumulation of evidence, and the theory has been revised continually to explain even more of the science of biology.

Directions: Based on what you have just read, fill in the circle that corresponds to the correct answer.

44. The theory of evolution is based on the idea that — 44 ① ② ③ ④ ⑤
(1) human beings descended from apes
(2) species have not changed over time
(3) complex forms of life have evolved from simpler forms
(4) only the strongest animals will survive
(5) organisms cannot adapt to different conditions

45. Darwin referred to natural selection as "survival of the fittest" because — 45 ① ② ③ ④ ⑤
(1) only finches with large beaks could compete for food on the Galapagos Island
(2) only living things able to compete successfully for food could survive
(3) populations multiply faster than food supplies
(4) evolving species could not interbreed
(5) only those living things that remain unchanged over long periods of time survive

46. Evolution is described in this passage as — 46 ① ② ③ ④ ⑤
(1) the study of structural similarities
(2) an outgrowth of studies in Social Darwinism
(3) a scientific theory attempting to explain the development of organisms over a long period of time
(4) the examination of prehistoric life through fossils
(5) a theory that explains why populations struggle over food supplies

For answers and explanations, see page 116.

BIOLOGY EXERCISE 7

Directions: Read the following passage and answer the questions. Fill in the circle that corresponds to the correct answer.

The Wild "Man" of Borneo

According to British zoologist John MacKinnon, an adult male orangutan from Sumatra stays close to his mate and offspring, while the males of Borneo roam more freely and do not seem to be part of a close family unit. What could account for the differences in their behavior?

The two Indonesian islands of Borneo and Sumatra are the only places orangutans, which are the most intelligent of all land animals, live today. The Malay words *orang hutan* mean "man of the forest." Some scientists think that these intelligent animals may engage in a type of sexual behavior once thought to be uniquely human.

Biologists used to think that only humans engaged in nonreproductive sex. Zoologist John MacKinnon reports that the orangutans on Sumatra seem to copulate without regard to the female's fertile period. It turns out that, on Sumatra, a female may have good reasons for keeping a strong male nearby. One way to do this may be to offer him sex.

MacKinnon suggests that, because there are leopards on Sumatra that prey on the orangutans, the females and infants there need protection. They also need a male to fight off siamangs, another ape species that competes with them for food. These two threats, present on Sumatra and not on Borneo, make it advantageous for Sumatran males to stick around to defend their offspring.

In contrast, on the island of Borneo, nothing ties males down; they wander freely through feeding ranges occupied by females and their young. The older, dominant males copulate with any fertile, receptive female they may encounter. Younger males, whose overtures tend to be rejected by the females, often resort to rape.

Here, the "love 'em and leave 'em" strategy actually works to the advantage of the females and infants. Were a large male to stay within their range, he would become a competitor for food. On Borneo, says primatologist David Agee, "the presence of the male may worsen his offspring's chance of survival." Being a good father on Borneo means not sticking around.

47. Male orangutans in Borneo and Sumatra exhibit different fathering behaviors. This is an example of
(1) adaptation
(2) transmutation
(3) reproduction
(4) evolution
(5) accumulation

47 ① ② ③ ④ ⑤

48. From this article, we can conclude that the differing fathering behaviors of the orangutans on Borneo and Sumatra are each suited to

48 ① ② ③ ④ ⑤

(1) producing the greatest number of surviving offspring
(2) maintaining cohesive families
(3) satisfying female orangutans
(4) competing for food with siamangs
(5) maintaining more orangutans in Sumatra

49. What threats do orangutans face on the island of Borneo?

49 ① ② ③ ④ ⑤

(1) Leopards and siamangs
(2) Younger male orangutans
(3) Infertile female orangutans
(4) Inadequate food supplies
(5) Too many offspring

For answers and explanations, see page 116.

ECOLOGY AND ECOSYSTEMS

One Town's Tragedy

Hardly anyone noticed when fish were found floating in the water and shellfish frequently died. Two years later, sea gulls started dropping out of the sky. In 1952, several cats contracted "dancing disease" and danced convulsively until they dropped dead. By 1958, there were no cats to be found. Eventually, dogs and pigs went mad and died.

Fishermen continued to fish, and people continued to eat the fish. In 1956, a six-year-old girl was admitted to the hospital with symptoms of brain damage. Her younger sister and four members of a neighboring family were found to be suffering the same symptoms. Within a year, fifty-two patients were identified; twenty-one died. The officials at the local plastics factory did an environmental study but never released their results.

By 1975, 103 people had died, and 700 more had been verified as seriously and permanently damaged. Severe mental retardation, tremors, and limb deformities were a few of the symptoms.

This disturbing description is not from a science fiction novel; it is a true story of ecological and social damage suffered in Minamata, Japan. Minamata suffered the first occurrence of mercury poisoning through man-made environmental pollution. Mercury in industrial wastes was dumped into the ocean, polluting the feeding grounds of shellfish and other fish. The poisoned fish were then eaten by animals and humans.

Mercury must accumulate in cells before there is enough to become toxic. Thus smaller animals were affected first. They weighed less and needed less mercury to display the symptoms. Eventually, enough mercury accumulated in the brain cells of humans to cause damage. The symptoms of this tragedy were first noticed among small children.

Minamata is in Japan, but our country has its own problems with industrial wastes polluting the environment. Love Canal, New York, and Times Beach, Missouri, are two notable examples. In both places, humans seeking progress and economic growth have disrupted their own environment. Only in the past few years have we begun to discover the extent of the damage to the ecosystem and realized that we must now pay the consequences.

Ecosystems

An ecosystem is an interacting system of living organisms and their environment. No organism lives alone. Every living thing is dependent on other living things and nonliving matter. Changes are constantly occurring in the ecosystem, but most of these changes help to keep an ecosystem in balance.

With the increased interest in ecology, many scientists are concerned by our relationship to the environment. Humans are dependent upon the environment to supply their food but have not always shown respect for their interdependence with the ecosystem. Some scientists believe that we have done so much damage to our ecosystem that the entire structure is in danger of collapsing. Since we are dependent on our ecosystem, its collapse would leave us without the resources necessary for our own existence.

Directions: Based on what you have just read, fill in the circle that corresponds to the correct answer.

50. The tragedy in Minamata was caused by
(1) poisoned fish
(2) mercury poisoning
(3) plastics
(4) "dancing disease"
(5) mental retardation

50 ① ② ③ ④ ⑤

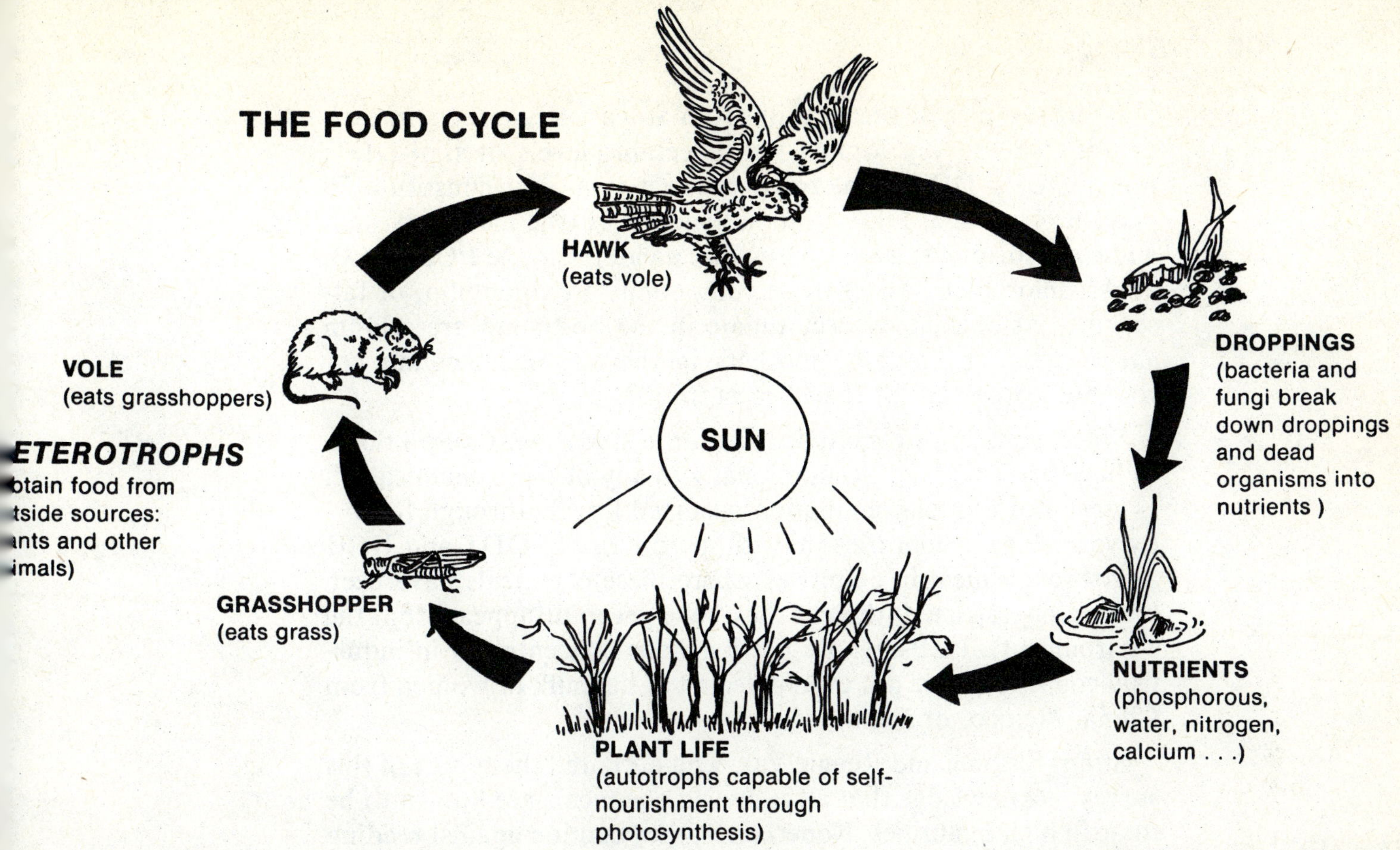

51. The diagram above shows one aspect of ecosystems—the food cycle. What repercussions would a decrease in the number of grasshoppers have on the cycle?

(1) An increase in the number of voles
(2) A decrease in the number of heterotrophs
(3) An increase in the number of heterotrophs
(4) An increase in the number of hawks
(5) No changes could be expected

51 ① ② ③ ④ ⑤

For answers and explanations, see page 116.

BIOLOGY EXERCISE 8

Directions: Read the passage below and answer the questions that follow. Fill in the circle that corresponds to the correct answer.

Nursing Mothers Share Pollutants

Breast-feeding mothers, particularly those in developing countries such as China, India, and Mexico, may be passing on the insecticides DDT, DDE, and HCH (hexachlorocyclohexane) to their babies in amounts that exceed "Acceptable Daily Intake" guidelines. In developed countries such as West Germany, the United States, and Japan, nursing mothers risk sharing contamination with PCBs. . . .

Using samples of breast milk from at least 50 mothers in each country, the survey looked for detectable levels of p,p^{1}-DDT; p,p^{1}-DDE (a DDT-breakdown product that may constitute 75 percent of residual DDT in milk; beta-HCH (the most persistent and toxic form of this systematic insecticide); and the PCB known as Arochlor 1260. All of these compounds are lipophilio, or fat-seeking. As such, they accumulate in the body and are seldom shed—except by nursing mothers in the fat that accounts for roughly 3.5 percent of their breast milk.

Since women were excluded from the study, who were known to have been exposed occupationally to any of these chemicals, it is presumed that contamination occurred largely through food—by vegetables in countries that still permit use of DDT and HCH and through meat in countries where these chemicals have been banned or restricted (though they have yet to disappear from the environment). PCBs, whose use has been concentrated in industrial countries, were not even detected in the milk of women from China, Mexico, or India.

Stuart Slorach and Reggie Vaz, who authored the report of this survey, acknowledge that all surveyed chemicals are known to be toxic in high quantities. Nonetheless, they caution against reading their findings as a reason to discourage breast-feeding. They point out that daily-intake guidelines have been developed as acceptable limits for a lifetime at that exposure, and a baby's contamination via breast milk usually lasts no more than a few months.

52. How do DDT and HCH enter the environment? 52 ① ② ③ ④ ⑤

(1) They are used by farmers as fertilizers.
(2) Women use them to prevent nausea while pregnant.
(3) They are by-products of the petrochemical industry.
(4) Farmers use them to protect their crops.
(5) They are car emissions.

53. Why were women known to have been exposed occupationally to DDT and HCH excluded from the study (lines 14–15)? 53 ① ② ③ ④ ⑤

(1) They could have been exposed to DDT and HCH at their jobs rather than through food sources.
(2) They were so weak from the high concentrations of DDT and HCH that the tests would have killed them.

(3) Their food sources had DDT and HCH levels higher than the "Accepted Daily Intake."
(4) They were not old enough to have the levels of DDT and HCH that the researchers were looking for.
(5) The levels of HCH and DDT in their breast milk would have been too high to measure.

54. Why were PCB levels higher in the breast milk of mothers from the United States and Japan than in those mothers from China and India?

54 ① ② ③ ④ ⑤

(1) Mothers in underdeveloped nations breast-feed more often than mothers in industrialized countries.
(2) PCBs have been used more in the underdeveloped nations.
(3) PCBs were used only in meat production in underdeveloped countries, and these women probably could not afford meat.
(4) There was so much DDT and HCH that the level of PCBs could not be measured.
(5) The use of PCBs is much more widespread in developed countries.

For answers and explanations, see page 116.

FLOWERING PLANTS

Plant Structure

When you look at a flower, you can see only a part of the plant. The part you cannot see is underground. A flowering plant has four basic parts: the roots (the part that is underground), the stem, the leaves, and the flower.

After a seed is planted in the ground, the roots are the first part of the plant to develop. Roots anchor plants and absorb nutrients from the soil. A plant may develop one of two types of root systems. Plants with fibrous roots, such as alfalfa, clover, and other grasses, have several slender roots. Other plants, such as radishes, carrots, and turnips, develop a taproot system with a primary root and many secondary roots as in the illustration on the next page.

Roots are generally covered with threadlike structures called root hairs, which aid in the absorption of nutrients from the soil.

There are two basic types of stems. One type is the kind that sticks out of the ground. Some stems are quite small, such as lettuce and small flowers, while others are quite large, such as the trunks and branches of trees. The other type of stem is found underground, such as the bulbs of tulips and daffodils or the tubers of potatoes. Each type of stem supports the leaves and flowers of the plant and carries nutrients from the roots of the plant to the other parts. At the top of the stem is a terminal bud from which the plant grows taller. There may also be lateral buds located along the stem that can develop into branches, leaves, or flowers.

The leaves are attached to the stem by a petiole. Many flowers have small leaflike structures where the petiole meets the stem. The leaf, or blade as it is sometimes called, has a network of veins that support the leaf and distribute the nutrients. The chief function of the leaf is the manufacture of food through photosynthesis.

THE PARTS OF A FLOWERING PLANT

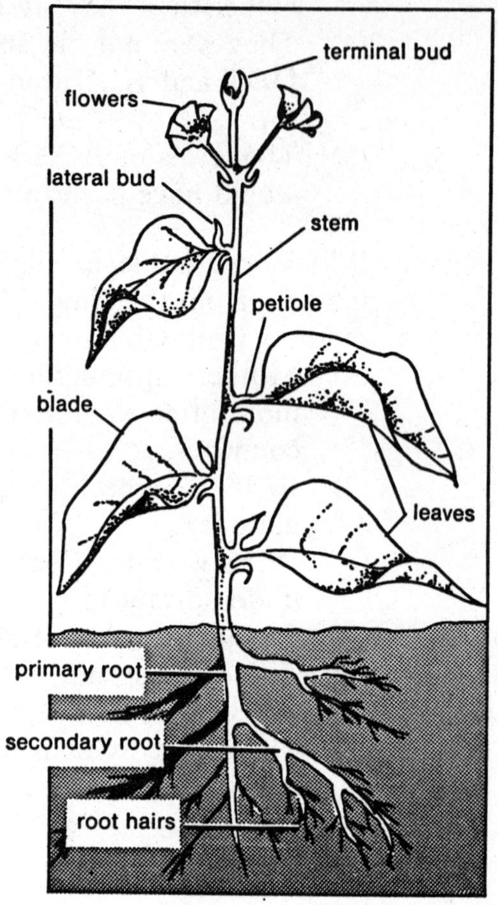

Pollination

Follow the diagram below as you read about reproduction in flowering plants.

Flowering plants reproduce through pollination. Pollen may be carried by bees or other insects, by wind, or by animals. Some plants pollinate themselves. Examples of this are cotton, peas, and tomatoes. In these plants, the stamen, the male reproductive organ of the plant, contains an ovary in its base. Pollination occurs when the pollen is carried from the anther at the top of the stamen to the stigma, at the top of the pistil. When the pollen is deposited on the stigma, it pushes down through the style into the ovary, where an egg cell is fertilized. The egg cell develops into a seed, which matures and falls to the ground. Eventually the seed will germinate into a young plant, completing the reproductive cycle.

THE PARTS OF A FLOWERING PLANT

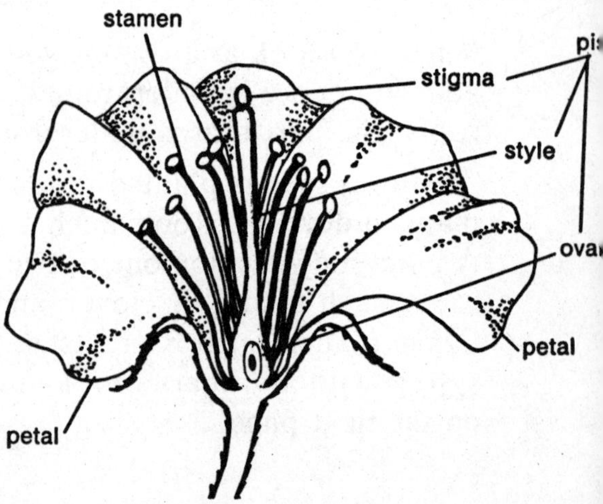

Directions: Based on what you have just read, fill in the circle that corresponds to the correct answer.

55. Mature oak trees frequently die four or five years after a building is constructed near them. Which of the following is the most likely reason for this? 55 ① ② ③ ④ ⑤
(1) The tree is no longer capable of reproducing.
(2) The supply of chlorophyll has been reduced.
(3) The delicate root structures have become severely damaged.
(4) The trees are unable to tolerate bitter winters.
(5) The trees have grown too tall.

56. According to the diagram of the pistil, the female reproductive organ of the flower has three main components. They are the 56 ① ② ③ ④ ⑤
(1) stigma, style, and ovary
(2) petal, stamen, and style
(3) petal, anther, and stigma
(4) pistil, ovary, and stamen
(5) anther, filament, and ovary

57. The process of reproduction is not complete until the 57 ① ② ③ ④ ⑤
(1) pollen is deposited on the stigma
(2) pollen enters the ovary
(3) egg cell is fertilized
(4) seed germinates
(5) seedling begins to mature

For answers and explanations, see page 117.

BIOLOGY EXERCISE 9

Directions: Read the following passage and answer the questions. Fill in the circle that corresponds to the correct answer.

Plant Nourishment

Plants make their own food by carrying out a series of chemical reactions called *photosynthesis*. This process depends on sunlight, water, air, and minerals. The key substance in the process of photosynthesis is chlorophyll, the greenish pigment that gives a plant its color. Only organisms that have chlorophyll are capable of photosynthesis.

Chlorophyll has the ability to trap the energy of sunlight, which is used to split molecules of water into their component elements of hydrogen and oxygen. The plant combines the hydrogen from the water molecule with carbon dioxide that has been absorbed through small openings called stomata and converts them into sugars and starches. The plant uses these sugars and starches as a source of energy and combines them with minerals such as potassium, phosphates, and sulfur to build plant tissue.

Photosynthesis creates water and oxygen as by-products. The plant releases the oxygen and water into the atmosphere through the stomata. Green plants provide most of the oxygen humans and animals need to survive. Air would become unbreathable if plants did not continuously replenish it.

58. What is photosynthesis?
(1) The reproductive cycle of flowers
(2) The food-making process in plants
(3) The growth of young plants
(4) The manufacture of chlorophyll
(5) The spreading of a plant's root structure

59. The absorption of carbon dioxide and release of oxygen in photosynthesis helps to aid which of the following processes in a human?
(1) Respiration
(2) Digestion
(3) Reproduction
(4) Circulation
(5) Elimination

60. When carbon dioxide enters a plant's leaves, it combines with hydrogen. These are converted to
(1) sunlight and water
(2) potassium, sulfur, and phosphorus
(3) photosynthesis
(4) stomata
(5) sugars and starches

61. Based on what you have read about photosynthesis, a mushroom
(1) is always poisonous
(2) produces its own food
(3) does not need very much water to grow
(4) has an extensive taproot system
(5) is not capable of producing its own food

For answers and explanations, see page 117.

ANSWERS AND EXPLANATIONS

INTRODUCTION TO BIOLOGY

Main Idea **1.** (4) While the other four choices are a part of the study of biology, only choice (4) gives the general definition of the field of study.

Finding Details **2.** (1) The last sentence of the first paragraph states that biology is divided into two broad divisions: botany, the study of plants, and zoology, the study of animals.

Drawing Conclusions **3.** (2) The essay points out that the invention of the microscope led to a discovery that revolutionized biology: the discovery of cells by Robert Hooke.

BIOLOGY EXERCISE 1

Making Inferences **4.** (5) On lines 4–8, the passage points out that the discovery of microbes was important to the study of disease, particularly in Pasteur's research on bacteria. Before the invention of the microscope, it would have been impossible to examine these tiny organisms.

Applying Ideas **5.** (3) The passage states that immunization is protection against the spread of bacteria. Rabies shots are an example of protection against disease.

Finding Details **6.** (3) The last paragraph describes Pasteur's development of the process named for him: pasteurization.

SYSTEMS OF THE HUMAN BODY

Making Inferences **7.** (1) While the epiglottis permits air to enter the windpipe, it closes when a person swallows food, preventing choking. Thus, you could infer that the epiglottis is essential to eating and digestion.

Finding Details **8.** (2) In the section on the digestive system, the passage states that food is absorbed by the blood and lymph vessels in the small intestine.

Applying Ideas **9.** (5) Involuntary muscles do their work without a conscious decision being made. The only choice that reflects the activity of an involuntary muscle is choice (5), a beating heart.

Drawing Conclusions **10.** (3) The section of the essay on the respiratory system explains that air entering the nose is cleaned by hair and the mucous lining of the nose. If you breathed through your mouth, the air would not be cleaned properly.

Finding Details **11.** (2) The skeletal structure and the muscles are highly interdependent. They work cooperatively, not independently.

Drawing Conclusions **12.** (4) The circulatory system supplies blood and oxygen to each of the systems of the body. Blockage of an artery would affect the balance between all these systems and undermine the well-being of the entire body.

BIOLOGY EXERCISE 2

Drawing Conclusions **13.** (5) Circulatory and heart problems are linked to lifestyle. Each individual has control over his or her lifestyle, including these factors.

Locating Details	**14.**	(1)	According to the information in lines 21–22, smoking can be a factor in hypertension since nicotine constricts the blood vessels.
Locating Details	**15.**	(5)	Blood pressure usually rises with age because the arteries become less elastic, slowing down the flow of blood.

NUTRITION

Finding Details	**16.**	(4)	The actual amount of food required and the number of calories needed depends on an individual's physical characteristics and level of activity.
Finding Details	**17.**	(2)	"Lone wolf calories" are calories that provide energy but no nutrients.
Reading Charts	**18.**	(3)	Look under the heading labeled "Fruits and Vegetables." Vitamins A and C are both nutrients gained from fruits and vegetables.
Reading Charts	**19.**	(3)	Of the groups listed, nursing mothers require the most servings of milk: 4 servings daily.
Reading Charts	**20.**	(5)	Both the fruit and vegetable and the milk group are listed as major sources of Vitamin A.
Applying Ideas	**21.**	(3)	If you look for Vitamin C on the chart, you'll find it in the Fruits and Vegetables row.

BIOLOGY EXERCISE 3

Main Idea	**22.**	(5)	The passage states that even moderate drinking (1) and (3) may be hazardous to the fetus.
Finding Details	**23.**	(1)	Mental retardation becomes evident as the FAS baby grows older. The other symptoms become apparent while the child is still an infant.
Applying Ideas	**24.**	(2)	According to the article, alcohol is poisonous to a fetus. If a woman finds that she has been pregnant for the past two months, it would be best for her to abstain from drinking for the rest of her pregnancy.

CELL STRUCTURE AND REPRODUCTION

Drawing Conclusions	**25.**	(5)	The passage states that there is a division of labor among cells in complex organisms. This means that the cells in complex organisms perform a more specialized function.
Finding Details	**26.**	(4)	The passage states that plant cells have two specialized structures not found in animal cells: cell walls and chloroplasts. These are also shown in the diagram.
Finding Details	**27.**	(4)	According to the first paragraph, protoplasm is the "substance of life," the essential living matter of plant and animal cells.

BIOLOGY EXERCISE 4

Finding Details	**28.**	(4)	In lines 3–4, the passage says that people with the RH factor are said to be RH-positive.
Drawing Conclusions	**29.**	(1)	If an RH-negative mother had already given birth to an RH-positive child, her blood would have anti-RH antibodies that could endanger the life of the second child.

Applying Ideas **30.** (1) According to the passage, RH-positive babies of RH-negative mothers have received transfusions to prevent death from RH incompatibility.

GENETICS

Finding Details **31.** (1) During each division, the chromosomes in each cell duplicate themselves so that each new cell contains chromosomes identical to those of the original.

Finding Details **32.** (2) According to the second paragraph, chromosomes are found in the cell nucleus.

Drawing Conclusions **33.** (2) The drawing shows that all of the first-generation offspring of a short plant and a tall plant are tall. This means that the trait for short height (t) is recessive.

BIOLOGY EXERCISE 5

Finding Details **34.** (5) According to lines 3–6, the hemoglobin in sickle cells releases too much oxygen.

Reading Charts **35.** (3) According to the diagram, there are four possibilities for a child's hemoglobin if one parent has sickle-cell trait and the other has sickle-cell anemia. Two of those four possibilities are that the child will have sickle-cell anemia.

Reading Charts **36.** (4) If one parent has normal hemoglobin and the other has sickle-cell anemia, all of the children will have sickle-cell trait. This fact is illustrated in the diagram.

Reading Charts **37.** (2) The diagram indicates that a person having one gene with normal hemoglobin and one gene with sickle-cell hemoglobin will have the sickle-cell trait.

MICROBES AND MICROBIOLOGY

Main Idea **38.** (4) Microorganisms are defined in the first sentence of the essay. Bacteria (5) are only one type of microorganism.

Applying Ideas **39.** (4) Although microoganisms do cause disease, choice (1), you cannot generalize that they are mainly harmful. Their ability to promote decay aids nature's processes and is useful in manufacturing, for instance, in raising bread and fermenting alcohol.

BIOLOGY EXERCISE 6

Finding Details **40.** (4) Influenza is spread by a virus. See lines 2 and 23.

Finding Details **41.** (5) Generally speaking, the elderly and those with lung diseases are considered to have a high risk of contracting fatal pneumonia when they catch the flu.

Finding Details **42.** (5) The article states (lines 14–18) that live virus vaccines carry the small risk that they will go back to a virile form and cause disease.

Drawing Conclusions **43.** (1) The last paragraph of the article states that the viruses seem to grow best in cells that line the nose, throat, and lungs—the respiratory system.

THE THEORY OF EVOLUTION

Main Idea	**44.**	(3)	One of the central ideas behind the theory of evolution is the proposition that more complex forms of life have evolved from simpler forms.
Drawing Conclusions	**45.**	(2)	According to Darwin's theories, as new species were formed, only those that were the best suited to their environment could survive competition for food supplies. The weak perished. This process is called *natural selection* or *survival of the fittest*.
Drawing Conclusions	**46.**	(3)	Evolution is a scientific theory that attempts to explain the development of organisms into higher forms. It does include the study of structural similarities, choice (1); examination of fossils, choice (4); and natural selection, choice (5); but these represent only segments of this theory.

BIOLOGY EXERCISE 7

Applying Ideas	**47.**	(1)	Darwin maintained that successful adaptation increases an animal's ability to survive and reproduce. In both cases, the fathering patterns adapted in such a way as to promote the survival of the species.
Drawing Conclusions	**48.**	(1)	The fathering behavior of orangutans on each of the two islands is best suited to ensuring the preservation of the species through the survival of the offspring.
Making Inferences	**49.**	(4)	The major threat to orangutans on Borneo is inadequate food supplies. While individual females may be threatened by young male orangutans, choice (2), this activity does not threaten the species. Leopards and siamangs, choice (1), are a threat only on Sumatra.

ECOLOGY AND ECOSYSTEMS

Finding Details	**50.**	(2)	The cause of the tragedy in Minamata was mercury poisoning from industrial waste.
Drawing Conclusions	**51.**	(2)	A decrease in grasshoppers would mean a reduction in the food supply available to the voles. This would eventually lead to a reduction in the number of voles. A decrease in the number of voles would reduce the food supply available to the hawks, eventually reducing the number of hawks. According to the diagram, grasshoppers, voles, and hawks are heterotrophs.

BIOLOGY EXERCISE 8

Making Inferences	**52.**	(4)	Line 3 mentions that HCH and DDT are both insecticides which are used to protect crops from insects.
Drawing Conclusions	**53.**	(1)	For the purposes of this study, women who had been exposed to DDT and HCH on their jobs had to be excluded since the researchers wanted to determine only how much DDT and HCH came from food supplies.
Making Inferences	**54.**	(5)	Lines 5–7 state that contamination from PCBs was a risk in developed countries such as West Germany, the United States, and Japan. From this statement you can infer that the use of PCBs has been much more widespread in developed countries. PCBs are an industrial waste from the manufacture of electrical equipment. You could also infer this from lines 18–20.

FLOWERING PLANTS

Making Inferences	**55.**	(3)	You can assume that the construction of a building might damage a tree's root structure. If the roots are damaged, the oak tree will not be able to sustain itself and will eventually die.
Reading Diagrams	**56.**	(1)	The arrows on the diagram indicate that there are three main parts of the pistil: the stigma, style, and ovary.
Finding Details	**57.**	(4)	The last paragraph of this essay states that the process of reproduction is not complete until the seed germinates into a young plant.

BIOLOGY EXERCISE 9

Main Idea	**58.**	(2)	The passage describes photosynthesis as the use of chlorophyll by plants to manufacture the food they need, using the nutrients brought up from the soil, energy from sunlight, and carbon dioxide drawn from the air.
Applying Ideas	**59.**	(1)	Photosynthesis in plants and respiration in humans are dependent on each other. The oxygen released by the plants is taken in by humans as a part of the respiratory process—breathing. According to lines 18–19, air would become unbreathable if plants did not replenish the oxygen supply.
Finding Details	**60.**	(5)	According to the passage, photosynthesis produces sugars and starches, which it combines with minerals to build tissue.
Drawing Conclusions	**61.**	(5)	A mushroom is not a green plant. Therefore, it does not contain chlorophyll and is not capable of producing its own food.

Earth Science

INTRODUCTION TO EARTH SCIENCE

The earth science passages on the GED science test are drawn from four areas: geology, meteorology, oceanography, and astronomy. Each of these broad areas contains many topics, but for the purposes of the GED test we will touch on only a few.

Geology is the study of the earth. The science of geology explains how the earth was formed and how it changes, as told through its mountains, mineral deposits, rocks and rock formations, and other structural features. Economic and environmental geology are two practical applications of this science. *Economic geology* is concerned with coal, metals, petroleum, and minerals that are of economic importance. Geologists are very active in the search for oil and other energy sources. *Environmental geology* applies the basic knowledge of geology to the study of environmental problems.

Meteorology is the study of the earth's atmosphere, the covering of air that surrounds the earth. Developments in meteorology have had a great effect on our ability to predict the weather. The modern observation of weather actually began in 1593 with the invention of the thermometer by Galileo. Forecasting was made possible with the advent of weather maps in the 1800s. Today, weather satellites equipped with cameras give us the ability to predict weather with greater accuracy.

Oceanography is the study of the earth's oceans. Oceanography is related to meteorology because the ocean's surface and the atmosphere affect each other. Scientists expect the study of oceanography to provide some important solutions to problems faced by humankind. The knowledge of sea plants and animals may increase our food supply. Oceanographers are studying how pollutants dumped into the sea affect the life cycles of sea animals and plants. They are also considering what must be done before more environmental damage occurs. Oceanographers study the sea bottom, looking for deposits of oil and natural gas to enlarge our supply of energy.

Astronomy, one of the oldest sciences, is the study of the sun, moon, planets, and stars. Astronomers study the movements and composition of the sun, other stars, and the planets. With their sophisticated instruments, astronomers are seeking clues about the origin and structure of our universe. Astronomers also study some of the effects of the heavenly bodies on the earth, such as radio communication interference caused by sunspots. With people now venturing into space with space shuttles and satellites, astronomy promises to play an even larger role in our lives.

Directions: Based on what you have just read, fill in the circle that corresponds to the correct answer.

1. You have just bought a large tract of land in an area known for its oil production. To determine if there is oil on your property you would hire a(n)
(1) meteorologist
(2) oceanographer
(3) geologist
(4) astronomer
(5) astronaut

1 ① ② ③ ④ ⑤

2. In the twentieth century, the practical application of our knowledge of meteorology has resulted in
(1) fewer meteors landing on the earth
(2) the ability to control damaging windstorms
(3) the invention of the thermometer
(4) humidity, precipitation, and air pressure
(5) an improved ability to predict the weather

2 ① ② ③ ④ ⑤

3. If you were planning a picnic for your annual family reunion, which of the following scientists would you be most interested in talking to?
(1) Astronomer
(2) Meteorologist
(3) Oceanographer
(4) Structural geologist
(5) Economic geologist

3 ① ② ③ ④ ⑤

4. Which of the following would be of greatest interest to an environmental geologist? 4 ① ② ③ ④ ⑤
(1) Elimination of an industrial waste dump
(2) The discovery of natural gas reserves
(3) Construction of a new telescope
(4) Prediction of earthquakes
(5) The discovery of a new star

5. Astronomy is the study of 5 ① ② ③ ④ ⑤
(1) the oceans and their currents
(2) fossils
(3) weather predictions
(4) the sun, moon, planets, and stars
(5) wind and air pressure

For answers and explanations, see page 133.

GEOLOGY

KEY WORDS

crustal plate—a part of the earth's surface that behaves as a rigid unit. Seven major plates that correspond roughly to the continents make up most of the earth's crust.

igneous—crystalline or glassy rock that is solidified from molten rock material

metamorphic—rock formed by the effects of pressure and high temperature on sedimentary or igneous rock

sedimentary—rock formed from loose material that solidifies over time

The Structure of the Earth

Most people assume that the earth is solid. They look at the surface of the earth and surmise that, since the surface is solid, the inner earth must be solid. The earth is actually covered by a very thin crust that is only about twenty miles thick. By studying the effects of earthquakes, scientists have learned that the earth beneath the crust is divided into three shells.

Below the crust lies the mantle, a thick layer of very dense rock that is about 1,800 miles thick. Close to the crust, the mantle has a temperature of about 1,600°F, while the area where the mantle meets the outer core is about 4,000°F.

EARTH'S STRUCTURE

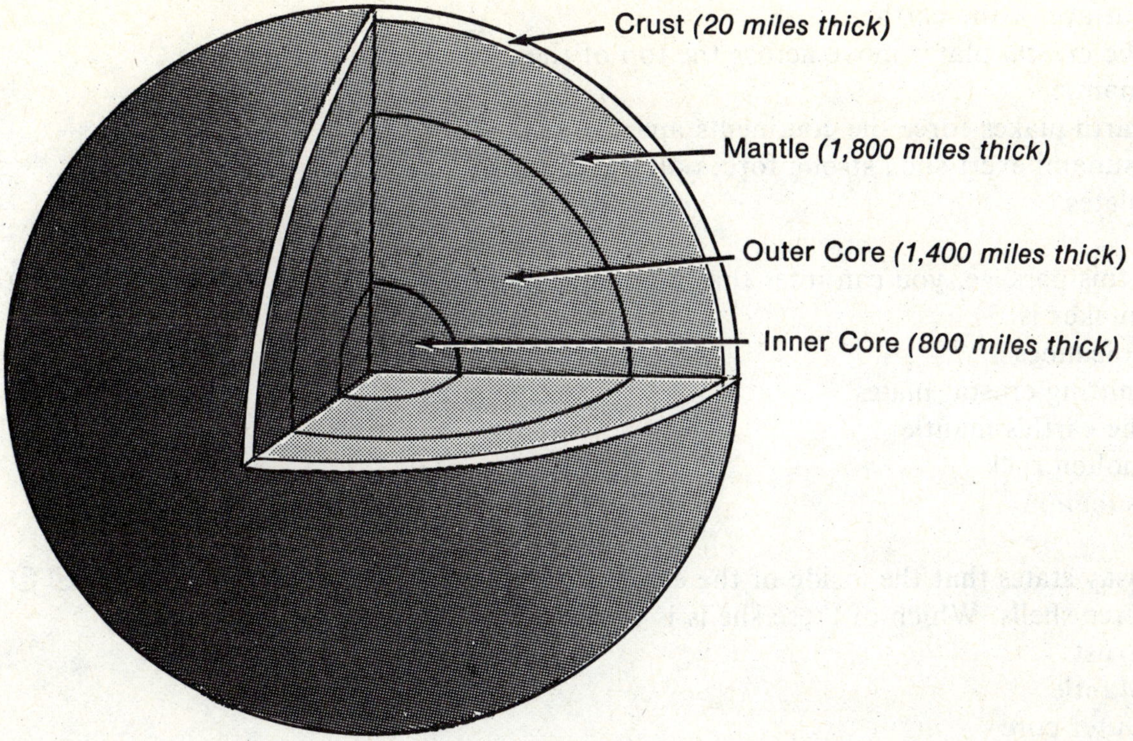

The outer core is about 1,400 miles thick and is made up of melted iron and nickel, with temperatures ranging from 4,000° F to 9,000° F. The inner core extends about 800 more miles and also consists of iron and nickel, but it is in a solid state due to the great pressures at that depth.

Because of the tremendous forces within the earth, the crust is subject to intense pressures. The plates that make up the crust are continually moving very slowly across the top of the mantle. The **crustal plates** move rigidly, carrying the continents along with them. Many scientists believe that the continents were once connected but were forced apart thousands of years ago. The movement of the crustal plates and the continents on them is frequently called continental drift.

Earthquakes are the result of the crustal plates' pushing against each other, causing the sudden release of energy. The vibrations of an earthquake can be slight or can cause considerable damage. Often, the greater damage results from the other catastrophic events triggered by earthquakes. For example, underwater earthquakes can cause tsunami, deadly tidal waves that can be sixty feet high.

Directions: Based on what you have just read, fill in the circle that corresponds to the correct answer.

6. The movement of the continents occurs because 6 ① ② ③ ④ ⑤
   (1) the earth's mantle carries the continents
   (2) the inner core moves, forcing molten lava to the surface of the earth
   (3) the crustal plates move across the top of the mantle
   (4) earthquakes force the continents apart
   (5) tsunami exert such strong forces on the crustal plates

7. From this passage, you can infer that a major cause of earthquakes is 7 ① ② ③ ④ ⑤
   (1) volcanoes
   (2) shifting crustal plates
   (3) the earth's mantle
   (4) molten rock
   (5) tsunami

8. The essay states that the inside of the earth is divided into three shells. Which of these shells is the thickest? 8 ① ② ③ ④ ⑤
   (1) Crust
   (2) Mantle
   (3) Outer core
   (4) Inner core
   (5) None of the above

For answers and explanations, see page 133.

Types of Rocks and Their Formation

The earth's crust contains three different types of rocks: **sedimentary**, such as sandstone; **igneous**, such as granite; and **metamorphic**, such as marble.

Sedimentary rocks, representing about 80 percent of the rock that lies under the earth's surface, were formed by sediments of small particles of mineral or organic matter. Most of these deposits occurred on the ocean floor. As time passed, the loose material changed into solid rock. Some types of sedimentary rocks were built up from the shells and skeletons of small shellfish, while other types were built up from the deposits of minerals that were dissolved in water. The third type of sedimentary rock was formed through the erosion of other rocks. Sedimentary rocks usually lie in horizontal strata, but they may fold or undergo other changes if intense pressures are placed on them.

Igneous rock forms when molten, or melted, rock material from deep within the earth pushes up to fill cracks formed by the movements of the crust. Through sheer heat and pressure, the molten material may also weaken the rock above it until the rock gives way to its force.

Metamorphic rock results when intense heat and pressure are applied to igneous or sedimentary rock. Pressure can force rock layers to fold and form mountains. Intense pressures and heat from molten rock can change any type of rock to metamorphic.

Directions: Based on what you have just read, fill in the circle that corresponds to the correct answer.

9. From the passage, we can conclude that **9** ① ② ③ ④ ⑤
(1) sedimentary rock is the most abundant type of rock
(2) metamorphic rock is less abundant than igneous rock
(3) igneous rock is built up in horizontal layers
(4) sedimentary rock is formed under intense heat and pressure
(5) igneous rock cannot be found in mountains

10. What transforms sedimentary rock into metamorphic rock? **10** ① ② ③ ④ ⑤
(1) Erosion
(2) Deposits of sediments
(3) Pressures of mountain building
(4) Heat from molten rock
(5) Heat and pressure

For answers and explanations, see page 133.

EARTH SCIENCE EXERCISE 1

Directions: Read the following passage and answer the questions. Fill in the circle that corresponds to the correct answer.

Mount St. Helens

Mount St. Helens exploded on May 18, 1980, with a force greater than that of the explosion of the first atomic bomb. The top 1,300 feet of the mountain was blown off, causing an avalanche of more than 3,000,000,000 cubic feet of mud, ice, ash, and rock. The force of the blast knocked down trees as far as twelve miles away and sent plumes of ash more than a mile into the sky. About sixty people were killed.

Even though Mount St. Helens had shown no signs of seismic activity for 123 years, a report issued in 1978 indicated that, based upon its behavior for the past 1,500 years, Mount St. Helens should be considered a potential hazard.

The first seismic activity was noted on March 20, 1980, and within a week more than 100 small earthquakes were caused by the movement of molten rock under the ground. On March 27, 1980, there was an eruption that left a 200-foot-wide crater. For the next five weeks there were more minor earthquakes, but even with all of this activity nobody expected the massive explosion that occurred at 8:32 A.M. on May 18, 1980.

Even though the May 18th explosion had not been anticipated, sixteen other minor eruptions before and after it were accurately predicted. Seismographs were used to detect earthquakes caused by the movement of molten rock. Tiltmeters placed on the ground near the volcano detected swelling. Aerial infrared surveys were used to identify "hot spots" of potential activity. The information that was gathered at Mount St. Helens will help scientists predict volcanic eruptions in the future.

11. What caused the explosion of Mount St. Helens? 11 ① ② ③ ④ ⑤
(1) An avalanche on the mountain
(2) Pressure from molten rock
(3) A series of minor earthquakes
(4) Plumes of ash
(5) An atomic explosion

12. Which of the following is used to help detect the movement of molten rock? 12 ① ② ③ ④ ⑤
(1) Seismographs
(2) Tiltmeters
(3) Aerial infrared surveys
(4) Seismic activity
(5) Volcanic eruptions

13. To what extent did scientists anticipate the May 18 explosion at Mount St. Helens? 13 ① ② ③ ④ ⑤
(1) They were able to predict it accurately by using tiltmeters to detect the swelling of the volcano.
(2) They were able to predict it accurately by using aerial infrared surveys that monitored "hot spots."
(3) They knew it was coming because seismographs detected a series of minor earthquakes just prior to the eruption.
(4) Someone who saw the avalanches was able to warn people about the coming explosion.
(5) They did not expect an explosion of that size.

For answers and explanations, see page 133.

METEOROLOGY AND OCEANOGRAPHY

KEY WORDS

atmosphere—the gases surrounding the earth or any heavenly body

continent—a large, unbroken area of land

Meteorology

Meteorology is the study of the earth's atmosphere and the changes in it that produce our weather. By analyzing data that have been collected, meteorologists can predict weather conditions with a fair degree of accuracy.

The **atmosphere** that surrounds the earth makes life possible. The atmosphere supplies the air that we breathe and protects us from some of the sun's rays, which would otherwise be dangerous. The atmosphere also regulates extreme temperatures that, if unchecked, would make life on earth impossible.

Our atmosphere is made up of many gases. While nitrogen is the most abundant gas, the atmosphere also contains oxygen, which all life needs to survive. Another gas, carbon dioxide, is used by plants in the production of their own food. The atmosphere also contains water vapor, made up of hydrogen and oxygen, which condenses to form clouds.

The atmosphere has four layers: (1) the *troposphere*, where most of our weather takes place; (2) the *stratosphere*, which contains a layer of ozone to protect us from the sun's ultraviolet rays; (3) the *mesosphere*, characterized by a dramatic drop in temperature with a rising altitude; and (4) the *thermosphere*, which contains particles that reflect radio waves back to earth, making it possible for us to receive radio waves from beyond the horizon.

Weather conditions in the troposphere, the layer of atmosphere closest to the earth, depend on four things: temperature, moisture, air pressure, and wind.

The heat in the atmosphere comes from the sun, either directly from the sun's rays or indirectly from heat the earth has absorbed from the sun. The atmosphere absorbs the heat and prevents it from escaping into space. Moisture enters the atmosphere as water vapor, mainly through evaporation from the ocean. The more water vapor there is in the air, the higher the humidity. As air is cooled, the water vapor condenses, either in the form of clouds or as raindrops.

Air pressure is the weight of the atmosphere as felt on the earth. Warm air is lighter than cool air and puts less pressure on the earth. Wind is created by the movement of air from high-pressure areas to low-pressure areas.

LEVELS OF THE EARTH'S ATMOSPHERE

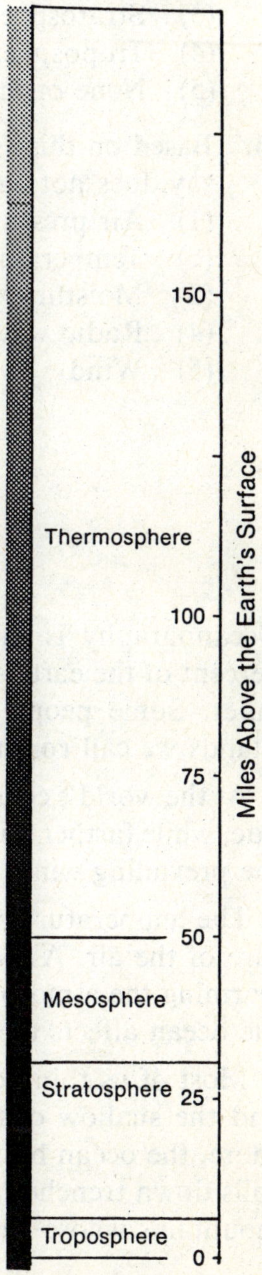

Directions: Based on what you have just read, fill in the circle that corresponds to the correct answer.

14. The most abundant gas in our atmosphere is
(1) carbon dioxide
(2) oxygen
(3) nitrogen
(4) water vapor
(5) helium

14 ① ② ③ ④ ⑤

15. Which layer of the atmosphere is closest to the earth's surface?
(1) Thermosphere
(2) Mesosphere
(3) Stratosphere
(4) Troposphere
(5) None of the above

15 ① ② ③ ④ ⑤

16. Based on this essay, which of the following would you say does not have an impact on weather conditions?
(1) Air pressure
(2) Temperature
(3) Moisture
(4) Radio waves
(5) Wind

16 ① ② ③ ④ ⑤

For answers and explanations, see page 134.

Oceanography

Oceanography is the study of the oceans. The oceans cover about 71 percent of the earth's surface and have an average depth of two and a half miles. Some people have called the world "one large sea," broken by islands we call **continents**.

At the world's coastlines, currents result from the rising and falling of the tide, while farther out, a worldwide system of currents operates that mimics the prevailing winds and influences our weather patterns.

The temperature of ocean water changes more slowly than the temperature of the air. As water moves past landmasses, it affects the climate by warming the atmosphere in the winter and cooling it in the summer. Thus the ocean affects the whole world's climate.

Most of us think of smooth sandy beaches when we think of the ocean and the shallow coastal waters in which we swim. But farther from the shore, the ocean bottom is anything but flat and smooth. The ocean floor rolls down trenches that are more than six miles deep and up underwater mountains whose peaks we know as islands.

The movement of the earth's crust forms ridges and deep trenches on the ocean floor. Scientists believe that, as the crust pulls apart, ridges are formed by soft, heated rock that rises to fill the opening. In areas where the moving crust pushes against other sections of the crust, deep trenches may form as portions slowly plunge back into the mantle. Most islands are the tops of undersea mountains, formed by volcanic activity.

Wave action is a major influence on our shorelines. The tides affect the development of beaches. Small waves put sand on the shore, while the larger waves characteristic of winter pull the fine sand off the beach, leaving pebbles and rocks. The oceans pound cliffs, breaking rocks into rubble. The ocean continues to erode the rubble until it is reduced to sand.

Directions: Based on what you have just read, fill in the circle that corresponds to the correct answer.

17. On a hot day, what kind of influence would you expect the ocean to have on the air temperature on land close to the ocean?

17 ① ② ③ ④ ⑤

(1) People would feel cooler when they went swimming.
(2) The ocean would raise the air temperature.
(3) The ocean would make people feel cooler since the ocean looks cool.
(4) The ocean would cool the air on land close to the ocean.
(5) The ocean would have no effect on the air temperature.

18. Ocean ridges are formed by

18 ① ② ③ ④ ⑤

(1) portions of the crust pushing into each other
(2) crustal plates that slowly plunge into the mantle
(3) underwater mountains
(4) erosion from wave action
(5) crustal plates pulling apart

19. From this essay, we could conclude that a beach experiencing hurricane tides for a period of time would become

19 ① ② ③ ④ ⑤

(1) mostly rocks and pebbles
(2) very fine sand
(3) overrun with marine animals
(4) a large expanse of sand
(5) nonexistent

For answers and explanations, see page 134.

EARTH SCIENCE EXERCISE 2

Directions: Read the passage below and answer the questions that follow. Fill in the circle that corresponds to the correct answer.

Acid Rain

In the 1950s, scientists noticed a rise in the acidity of lakes in Ontario, Canada, and several midwestern states. The first victims they observed were the eggs of fish and amphibians. Soon thereafter they noticed that frogs were also dying. As the waters became increasingly acidic, the natural cycle of plant decay was disrupted, meaning that nutrients were not being returned to the environment. There are now many lakes that have no fish because the water has become so acidic that it cannot support fish. What caused this? The answer is believed to be acid rain.

Acid rain forms from sulfur dioxide and nitrogen oxide. Sulfur dioxide is released into the environment during fuel combustion, as in the production of iron and the refining of oil. Nitrogen oxide is released through automobile and truck exhausts. When these two compounds combine with the moisture in clouds, a solution of nitric and sulfuric acids forms.

These emissions are dispersed and spread by the wind. Besides damaging lakes, the acids damage buildings, causing the surface of the stone to flake off. Acid rain may also cause changes in the soil, affecting both farmlands and forests, although scientists are still not sure about the ultimate effects of these changes.

Several solutions to the problem of acid rain have been tried. In Sweden, scientists are restoring lakes by dumping large quantities of lime into them to neutralize the acid. Even though the lake is "cleaned up," the process must be repeated as more acid rain falls into it. Other scientists are attempting to develop new, resistant strains of fish, but this approach seems likely to produce only limited success.

The only permanent solution is to stop the acidic emissions at their source. While the technology necessary to do so is expensive, it is important to remove sulfur dioxide from emissions before they are released into the environment. One method involves chemical "scrubbers" that absorb sulfur dioxide. Progress in curbing acid rain is being made, but more action must be taken so that our environment does not become irrevocably altered.

20. Sulfur dioxide and nitrogen oxide are spread **20** ① ② ③ ④ ⑤

(1) in groundwater
(2) by the wind
(3) through infected fish eggs
(4) by new strains of fish
(5) in the food supplies

21. Which of the following does the passage regard as a method to prevent acid rain?

21 ① ② ③ ④ ⑤

(1) Dump large quantities of lime into lakes.
(2) Develop strains of fish resistant to sulfur dioxide.
(3) Require catalytic converters on all automobiles.
(4) Cut back on the production of iron.
(5) Use chemical "scrubbers" that absorb sulfur dioxide.

For answers and explanations, see page 134.

ASTRONOMY

KEY WORDS

galaxy—a group of billions of stars
universe—the system of all the galaxies

People have always gazed into the skies with a sense of wonder, astonished by the vastness of the constellations. Astronomy, the oldest science, is the study of the movement and composition of the sun, moon, planets, and stars.

Some scientists believe that the **universe** may be as much as 15 billion years old; others feel that it is younger than that. Many theories have been developed to explain how the universe, and our solar system in particular, came into existence. One widely accepted theory holds that large, whirling clouds of gas and dust shrank and broke into individual units that became planets.

Our solar system is only one tiny part of a group of stars called the Milky Way Galaxy. Our **galaxy** consists of about 100 billion stars and is approximately 100,000 light years across, a light year being the distance light can travel in a year. (Light travels at about 186,000 miles per second.)

Our solar system consists of the sun and its family of nine planets, including the earth. Early observers believed that the earth was the center of the universe, and it was only 400 years ago that scientists came to understand that the nine planets revolve around the sun. We now know that the planets are held in their orbits around the sun by the sun's powerful gravitational force. Some of these planets have their own satellites, or moons, orbiting them like miniature solar systems. Our solar system also has thousands of small, planetlike asteroids that revolve around the sun between the orbits of Jupiter and Mars. Some scientists believe that these are the remains of a planet that was destroyed.

OUR SOLAR SYSTEM

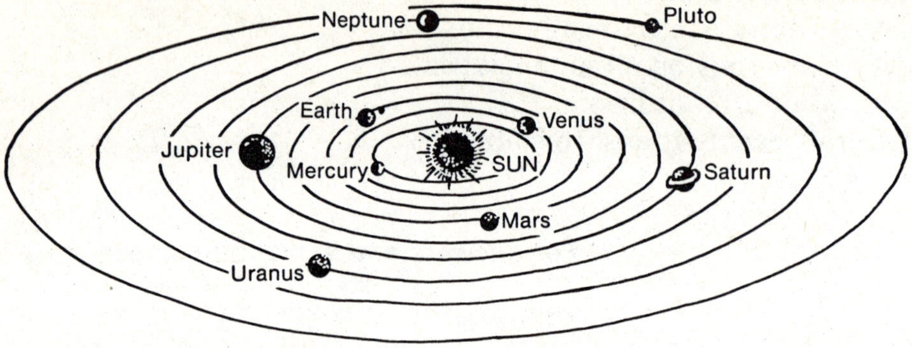

The smallest planets are the four closest to the sun: Mercury, Venus, Earth, and Mars. The largest planets are Jupiter; Saturn, which is famous for its rings; Uranus; and Neptune. Very little is known about Pluto. It has an eccentric orbit that sometimes overlaps Neptune's.

The sun is a star that radiates heat and light. Its energy is widely believed to come from nuclear reactions in its innermost regions. The planets shine only by the light that they reflect from the sun.

The earth makes one revolution around the sun each year, moving at a speed of about 67,000 miles per hour. Every 24 hours, the earth rotates on its axis. At any single moment, half of the earth is illuminated by the sun, while the other half lies in darkness.

TILT OF THE EARTH'S AXIS

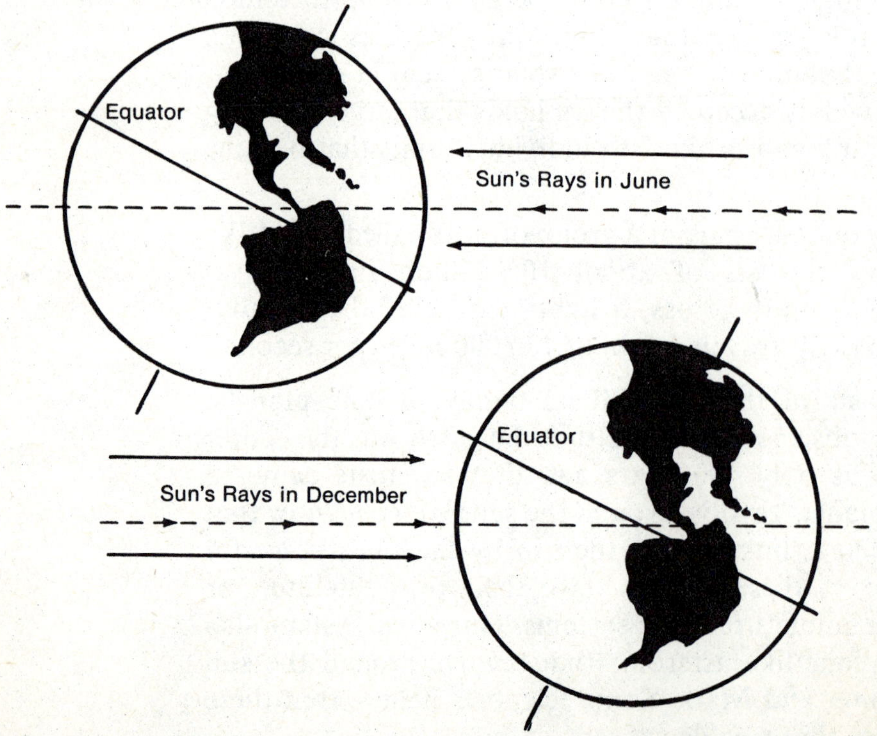

The earth's axis is an imaginary line that runs through the North and South Poles. It is the tilt in this axis, not the earth's distance from the sun, that is responsible for changes in the seasons. Winter occurs in the Northern Hemisphere when the northern half of the earth is tilted away from the sun and the sun's rays are less direct. The Northern Hemisphere has summer when that hemisphere is tilted toward the sun and receives sun rays that are more direct.

Directions: Based on what you have just read, fill in the circle that corresponds to the correct answer.

22. The essay refers to a natural phenomenon that holds the planets in predictable orbits. This force is called
(1) gravity
(2) rotation
(3) revolution
(4) orbit
(5) orientation

22 ① ② ③ ④ ⑤

23. Within a twenty-four-hour period, we have both day and night. Which of the following accounts for this?
(1) The sun moves around the earth once a day.
(2) The earth revolves around the sun each day.
(3) The earth rotates on its axis.
(4) The earth rotates each year and orbits each day.
(5) The sun orbits the earth in a twenty-four-hour cycle.

23 ① ② ③ ④ ⑤

24. The Milky Way Galaxy contains approximately
(1) 100,000 stars
(2) 100 million stars
(3) 100 billion stars
(4) 100,000 solar systems
(5) 100 million planets

24 ① ② ③ ④ ⑤

25. When the Northern Hemisphere is tilted toward the sun, what season is it for that half of the earth?
(1) Spring
(2) Summer
(3) Fall
(4) Winter
(5) None of the above

25 ① ② ③ ④ ⑤

For answers and explanations, see page 134.

EARTH SCIENCE EXERCISE 3

Directions: Read the passage below and answer the questions that follow. Fill in the circle that corresponds to the correct answer.

The Search for Extraterrestrial Intelligence

E.T., *Close Encounters of the Third Kind*, and H. G. Wells' *War of the Worlds* all suggest that many people believe in the possibility of intelligent life on other planets. Reports of UFOs (Unidentified Flying Objects) lend support to many people's belief that the earth is visited by creatures from outer space. The public's imagination is further fired up when respectable scientists begin to take UFO reports seriously. Although no one has offered a reasonable, scientific explanation for the phenomenon, millions remain convinced that UFOs may be of extraterrestrial origin.

On a broader scale, the question of whether life exists in outer space is one that scientists are taking very seriously. They argue that deep channels on Mars could have been riverbeds and that Mars may have once supported life.

Using advanced radio telescopes aimed into outer space and monitored continuously by computers, scientists are able to listen to about 130,000 radio channels simultaneously in the hope that there are intelligent beings attempting to communicate and make themselves known. Maybe someday they will pick up a message from an E.T. phoning home.

26. Based on the passage, you can conclude that the search for other forms of intelligence is

(1) a futile effort
(2) something that is done only in the movies
(3) done with very simple technology
(4) a waste of money
(5) a reasonable activity with much potential

26 ① ② ③ ④ ⑤

27. The passage indicates that the search for extraterrestrial intelligence is being made through

(1) manned space exploration
(2) works of fiction
(3) radio broadcasts from earth
(4) radio telescopes and computers
(5) civil defense units

27 ① ② ③ ④ ⑤

For answers and explanations, see page 135.

ANSWERS AND EXPLANATIONS

INTRODUCTION

Applying Ideas	**1.**	(3)	A geologist would be the scientist concerned with the earth's structure and composition.
Finding Details	**2.**	(5)	The paragraph on meteorology states that, in recent years, advances in meteorology have resulted in the ability to predict weather.
Applying Ideas	**3.**	(2)	If you were planning a picnic for your annual family reunion, you would probably be concerned about choosing a time when the weather would most likely be pleasant. The scientist who could help you most with that would be the meteorologist.
Drawing Conclusions	**4.**	(1)	An environmental geologist applies a knowledge of geology to environmental problems. Of the situations listed, the geologist would most likely be concerned with the dangers posed by an industrial waste dump.
Finding Details	**5.**	(4)	The last paragraph of the essay states that astronomy is the study of the sun, moon, planets and stars.

GEOLOGY

Finding Details	**6.**	(3)	The essay states that the crustal plates are continuously moving across the top of the mantle, carrying the continents along with them.
Finding Details	**7.**	(2)	The last paragraph states that earthquakes are the result of crustal plates' pushing against each other.
Finding Details	**8.**	(2)	The passage states that the mantle is about 1,800 miles thick. The outer core, choice (3), is 1,400 miles thick and the inner core, choice (4) is 800 miles thick. The crust, choice (1), is not considered one of the three shells.
Drawing Conclusions	**9.**	(1)	The passage states that 80 percent of the rock under the earth's surface is sedimentary rock.
Finding Details	**10.**	(5)	The last paragraph states that the heat from molten rock and tremendous pressure can change sedimentary rock to metamorphic rock. Choices (3) and (4) include only parts of the answer. Choices (1) and (2) deal with the formation of sedimentary rock, not how it is transformed into metamorphic rock.

EARTH SCIENCE EXERCISE 1

Making Inferences	**11.**	(2)	The passage indicates that the massive eruption was preceded by a series of minor earthquakes (3) but that these were caused by the movement of molten rock. Tiltmeters were used to detect swelling, indicating that something was pushing up from below the ground. Choices (1) and (4) are results of the eruption. Choice (5) is unreasonable.
Making Inferences	**12.**	(1)	The fourth paragraph indicates that seismographs are used to detect earthquakes caused by the movement of molten rock.

Drawing Conclusions **13.** (5) The last paragraph states that the explosion caught geologists by surprise, indicating that this was an event that they had been unable to predict. Choices (1), (2), and (3) correctly describe the functions of tiltmeters, aerial infrared surveys, and seismographs. However, the passage states that these tools helped geologists predict sixteen other minor earthquakes but not the huge eruption of May 18.

METEOROLOGY AND OCEANOGRAPHY

Finding Details **14.** (3) The passage specifically states that nitrogen is the most abundant gas in our atmosphere. All of the other gases—choices (1), (2), (4), and (5)—are present in lower quantities. Helium, choice (5), is not even mentioned in the passage.

Finding Details **15.** (4) The fifth paragraph of the passage states that the troposphere is the layer of atmosphere closest to the earth. You also could have inferred that, since the troposphere is where our weather takes place, it is the layer closest to the surface of the earth.

Drawing Conclusions **16.** (4) The last two paragraphs discuss factors that influence weather conditions. Radio waves, choice (4), are not mentioned.

Applying Ideas **17.** (4) The essay states that ocean water cools the surrounding climate in the summer and that a large body of water can be expected to moderate hot temperatures.

Finding Details **18.** (5) According to the passage, as the crust pulls apart, ocean ridges are formed by soft, heated rock that rises to fill the opening.

Drawing Conclusions **19.** (1) Large waves, such as those caused by hurricanes, pull the sand out to sea, leaving mostly rocks and pebbles. The beach would still exist but would consist of much coarser material. There is no mention of a beach's being overrun by marine animals, choice (3).

EARTH SCIENCE EXERCISE 2

Finding Details **20.** (2) Some pollutants are spread by the other methods, but, according to the third paragraph, sulfur dioxide and nitrogen oxide are dispersed by the wind.

Drawing Conclusions **21.** (5) Chemical "scrubbing" is presented as a method that would remove sulfur dioxide before it is released into the air. Choices (1) and (2) do not prevent acid rain. They are temporary solutions for the effects of acid rain. Choices (3) and (4) are not mentioned at all.

ASTRONOMY

Finding Details **22.** (1) The essay mentions that the planets are held in their orbits by gravitational forces. While the planets revolve around the sun, choices (3) and (4), and they rotate, choice (2), the force that holds them in their orbits is gravity.

Making Inferences **23.** (3) The essay states that the earth rotates on its axis every 24 hours. At any single moment, half of the earth is illuminated while the other half is in darkness. While ancient observers believed that the sun revolved around the earth, choices (1) and (5), scientists now understand this is not the case. The earth revolves around the sun once a year, disproving choices (2) and (4).

Finding Details	**24.**	(3)	The essay states that there are 100 billion stars in the Milky Way Galaxy.
Finding Details	**25.**	(2)	The last sentence in the essay states that in the summer the Northern Hemisphere is tilted toward the sun.

EARTH SCIENCE EXERCISE 3

Drawing Conclusions	**26.**	(5)	The second paragraph supports this conclusion. The passage states that scientists are taking the possibility of extraterrestrial life seriously and believe that Mars may have once supported life. This and other evidence motivates scientists to keep searching.
Finding Details	**27.**	(4)	The radio telescope is mentioned in line 15, and the passage says that these telescopes are constantly monitored by computers.

Chemistry

WHAT IS CHEMISTRY?

KEY WORDS

matter—anything that occupies space

hypothesis—an unproven assumption that is based on a set of facts or ideas

Chemistry is the study of the structure, composition, and properties of matter. **Matter** can be defined as anything that has weight and takes up space, such as water, air, a desk, or a person. Chemistry is concerned with the structure of water and the composition of a desk. You might say that a desk is made of metal. But what is the metal made of? What properties and qualities does the metal have? In order to find out, chemists try to break it down into its simplest form. By understanding the basic composition and structure of the simplest forms of matter, chemists are able to make scientific breakthroughs that affect us all.

For example, the science of chemistry has had a tremendous impact on the plastics industry. More and more of the things we use are made of plastics. Glass milk bottles have been replaced by plastic jugs. Your telephone, the floor tile in your home, and even the dashboard in your car are all made of types of plastics. The use of plastics is growing at a phenomenal rate, and chemistry is responsible for this exciting and useful development.

Chemistry is also responsible for the development and growth of the medicinal drug industry. Long ago, medicine consisted of simple remedies such as juices from plants and trees, extractions from the soil, and other products of nature. By studying the composition of these remedies, chemists were able to isolate the compounds that were responsible for soothing or pain-relieving effects and to develop these compounds into drugs.

The most common drug in the world is a good example of this process. During the seventeenth and eighteenth centuries, American Indians used the bark from a willow tree as a treatment for fever. Later, chemists discovered that the chemical that actually produced the pain-relieving effect was *salicylic acid*, and they began to produce this chemical for medicinal purposes. The drug was effective, but it had very unpleasant side effects, including severe irritation of the stomach lining. Further advances enabled chemists to reduce this irritation by adding another element to the salicylic acid. The resulting drug, called *acetylsalicylic acid*, is what we know today as aspirin.

Directions: Based upon what you have just read, fill in the circle that corresponds to the correct answer.

1. You can infer from this essay that chemistry — 1 ① ② ③ ④ ⑤
(1) deals only with the manufacture of drugs
(2) can be used only by doctors
(3) has not made any breakthroughs for years
(4) is a laboratory-based science that has little to do with everyday life
(5) was involved in the development of transparent plastics

2. The study of chemistry — 2 ① ② ③ ④ ⑤
(1) considers only the color and shape of matter
(2) focuses on medicines and their effects
(3) breaks down matter into its simplest form
(4) is based on the needs of the plastics industry
(5) has not had an impact on our daily lives

3. The drug that we call *aspirin* — 3 ① ② ③ ④ ⑤
(1) can be dug up from the soil
(2) still causes intense stomach irritation
(3) was accidentally invented by a chemist in his laboratory
(4) is a chemical found in nature and modified by chemists
(5) is called *salicylic acid* by scientists

For answers and explanations, see page 160.

CHEMISTRY EXERCISE 1

Directions: Read the passage below and answer the questions that follow. Fill in the circle that corresponds to the correct answer.

The Scientific Method

Chemistry evolved from alchemy—an art that was practiced for more than 2,000 years. Like scientists today, alchemists were interested in the composition of matter and why certain things happened. Unlike modern chemists, the alchemists relied on a combination of religion, trial-and-error experimentation, and philosophy to answer their questions.

The study of alchemy originated with observations from everyday life. Chemical reactions were observed in metal-working, glass and pottery making, and the creation of some medicines. These observations led some alchemists to believe that all matter was made up of earth, fire, water, and air. Other alchemists attempted to use their observations to convert nonprecious metals into gold and to develop a drug that would cure all illnesses.

Of course these attempts were unsuccessful, and the alchemists failed. However, through their failures, these early scientists began to see the need for systematic study. They realized that trial-and-error concoctions were not the result of an organized and productive means of investigation. What eventually developed is what scientists today call the *scientific method.*

1. The first step of the scientific method involves identifying what will be investigated. With the knowledge of prior scientific discoveries, the scientist must identify a specific problem to be solved.

2. The second step in this scientific method is observation and experimentation. Experiments must be performed and recorded with utmost care so that the results obtained are repeatable by other scientists at other times in other places.

3. The third step involves drawing conclusions. For instance, if a mixture of iron and yellow dyes continually fails to produce gold, it is the scientist's task to find out why. A chemist who repeatedly observes the same reaction between two elements in his laboratory must make a hypothesis, or an educated guess, to explain why these elements behave the way they do.

4. Then the hypothesis must be tested. Some alchemists believed that if copper were mixed with zinc, the resulting metal would be gold. They did get a metal that looked like gold, but further tests showed that they were wrong. They had actually

produced brass. This shows the importance of subjecting a hypothesis to strict testing so that a scientist's ideas can be verified or proven wrong. After repeated testing of a hypothesis, the scientist may summarize the results obtained and thereby propose a general scientific principle.

4. Which of the following represents the correct steps in the scientific method? 4 ① ② ③ ④ ⑤

(1) Identification of a problem, formation of a hypothesis, experimentation
(2) Choosing a problem, then making a series of educated guesses
(3) Identifying a theory, then trying to find out how to prove it
(4) Identifying a problem, then developing a formula to answer it
(5) Choosing a problem, experimenting, making a hypothesis, and testing the hypothesis

5. You can infer from the passage that the alchemists 5 ① ② ③ ④ ⑤

(1) were interested only in the money they could make
(2) invented the scientific method
(3) were more interested in the outcome of their work than in developing accurate methods
(4) discovered a way to make gold by using trial and error experimentation
(5) had no influence on modern chemistry other than developing the scientific method

For answers and explanations, see page 160.

ATOMS AND ATOMIC THEORY

KEY WORDS

electron—negatively charged particle orbiting the nucleus

proton—positively charged particle found in the nucleus of an atom

neutron—particle found in the nucleus that has no electrical charge

atomic weight—approximately equal to the sum of the number of protons and the number of neutrons in an atom

atomic number—the number of protons in an atom

Matter is made up of different elements, which are, in turn, made up of atoms. An atom is the smallest particle of an element that has the properties of that element. For example, if you were to divide a lump of iron into smaller and smaller pieces, until you could divide it no further, you would eventually have a single atom of iron.

Atoms are extremely small—so small that they cannot be seen, even under a microscope—but they are important to science. The building blocks of an atom are protons, neutrons, and electrons. Below is a diagram of an atom.

THE ATOM

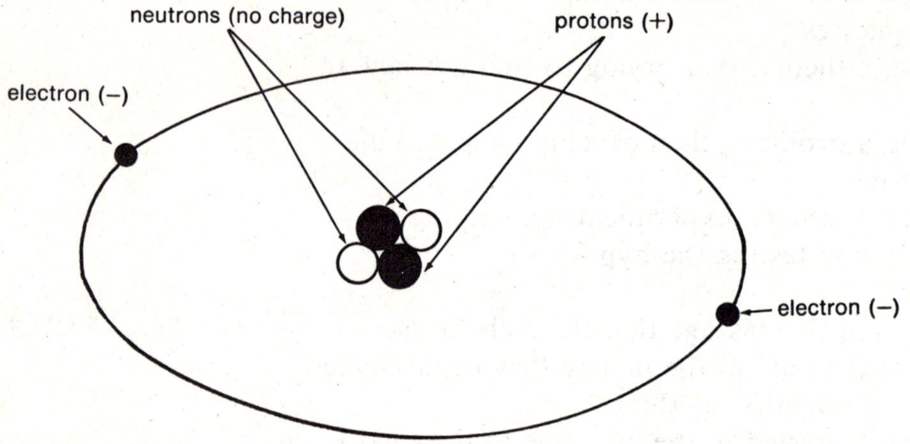

An **electron** is the smallest particle in an atom, and it has a negative electrical charge. It is useful to think of each atomic electron as revolving around the central nucleus in much the same way as the earth moves around the sun. The nucleus itself is made up of two types of larger particles, **protons** and **neutrons**. A proton has a positive electrical charge, while a neutron has no electrical charge at all. Although an atom may have many electrons, nearly all of its mass (material substance) consists of the mass of its nuclear protons and neutrons.

Although atoms contain electrically charged particles, all elements are perceived as electrically neutral, meaning that each atom has an equal number of protons (positively charged) and electrons (negatively charged). Scientists have also discovered that the proton and neutron have approximately the same weight, whereas the electron weighs considerably less. To simplify things, chemists often agree to give the proton and neutron each a weight of 1 and to assign the electron a weight of 0.

It is convenient to think of the **atomic weight** of an atom as the sum of the number of protons and neutrons in the atom. An atom of oxygen, with eight protons, eight neutrons, and eight electrons, would have an atomic weight of sixteen since the sum of the number of protons and neutrons is sixteen and the number of electrons is not included in the atomic weight.

A device that chemists use to classify an atom is its **atomic number**. The atomic number is equal to the number of protons in the atom and is used to categorize elements on a chart called the *periodic table*.

Directions: Based on what you have just read, fill in the circle that corresponds to the correct answer.

6. The smallest particle of an element that has all the properties of that element is called
(1) an electron
(2) an atom
(3) electrically charged
(4) a proton
(5) a nucleus

6 ① ② ③ ④ ⑤

7. Which of the following statements is true?
(1) The atomic number of an atom is equal to the number of protons.
(2) The weight of an electron is considered to be one.
(3) The weight of a proton is about the same as the weight of an electron.
(4) A proton revolves around the nucleus of an atom.
(5) A neutron has a positive charge.

7 ① ② ③ ④ ⑤

8. The element Helium has two electrons, two protons, and two neutrons. Its atomic number is
(1) six
(2) four
(3) two
(4) zero
(5) not measurable

8 ① ② ③ ④ ⑤

For answers and explanations, see page 160.

CHEMISTRY EXERCISE 2

Directions: Read the passage below and answer the questions that follow. Fill in the circle that corresponds to the correct answer.

The Periodic Table

What makes one atom different from another? The answer is that atoms differ in the number of protons in the nucleus and in the arrangement of electrons orbiting the nucleus. Today, scientists know of 108 elements, each having a different structure. Hydrogen is the simplest element, containing one electron that revolves around its single-proton nucleus. Magnesium, on the other hand, is more complex, having 12 electrons. These 12 electrons do not all follow the same path. As shown in the diagram on the next page, magnesium has different electron "shells," with varying numbers of electrons in each shell.

MAGNESIUM

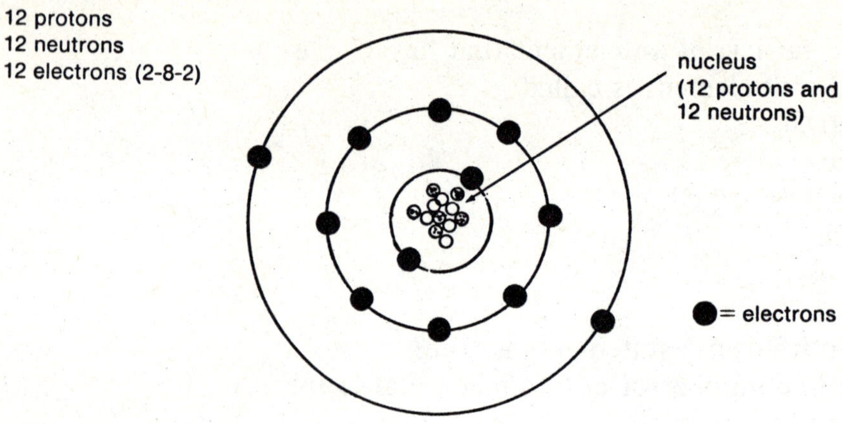

The differing shell structure of different elements is the basis for the design of the periodic table. The periodic table is a chart that organizes elements in rows and columns. The symbol for each of the 108 elements is shown with an atomic number (the number of protons). The elements in each "group" (vertical column) have the same number of electrons in their outermost shell. The elements in each "period" (horizontal row) have the same number of shells.

The periodic table was invented in 1869 by Dmitry Mendeleyev, a Russian chemist, after many others had attempted to organize the elements. Mendeleyev's work is particularly remarkable since, at the time of his invention, only 63 elements were known. He predicted the existence of other elements and left the appropriate spaces for them on his table.

Out of the 108 known elements, 89 are found in measurable amounts in nature, and the others have been created through experimentation.

9. The design of the periodic table indicates that — **9** (1) (2) (3) (4) (5)

(1) there are 63 elements in the universe
(2) the elements are unrelated and cannot be organized
(3) neutrons are the basis upon which elements are grouped
(4) all of the elements are found in nature
(5) elements can be grouped according to their electron structures

PERIODIC TABLE OF THE ELEMENTS

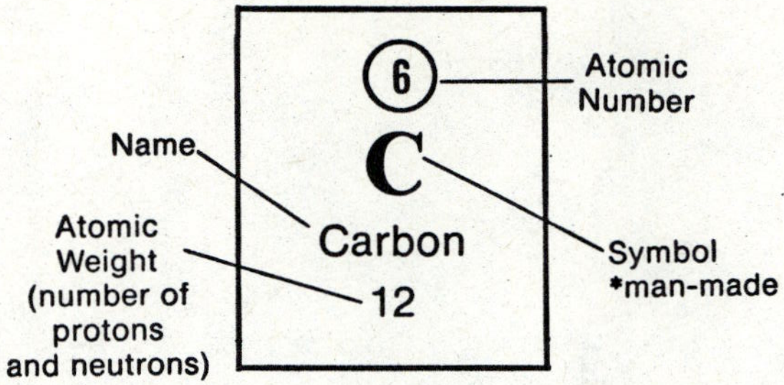

PERIOD																		
1	(1) H Hydrogen 1																	(2) He Helium 4
2	(3) Li Lithium 7	(4) Be Beryllium 9											(5) B Boron 11	(6) C Carbon 12	(7) N Nitrogen 14	(8) O Oxygen 16	(9) F Fluorine 19	(10) Ne Neon 20
3	(11) Na Sodium 23	(12) Mg Magnesium 24											(13) Al Aluminum 27	(14) Si Silicon 28	(15) P Phosphorus 31	(16) S Sulfur 32	(17) Cl Chlorine 35	(18) Ar Argon 40
4	(19) K Potassium 39	(20) Ca Calcium 40	(21) Sc Scandium 45	(22) Ti Titanium 48	(23) V Vanadium 51	(24) Cr Chromium 52	(25) Mn Manganese 55	(26) Fe Iron 56	(27) Co Cobalt 59	(28) Ni Nickel 59	(29) Cu Copper 64	(30) Zn Zinc 65	(31) Ga Gallium 70	(32) Ge Germanium 73	(33) As Arsenic 75	(34) Se Selenium 79	(35) Br Bromine 80	(36) Kr Krypton 84
5	(37) Rb Rubidium 85	(38) Sr Strontium 88	(39) Y Yttrium 89	(40) Zr Zirconium 91	(41) Nb Niobium 93	(42) Mo Molybdenum 96	(43) *Tc Technetium 99	(44) Ru Ruthenium 101	(45) Rh Rhodium 103	(46) Pd Palladium 106	(47) Ag Silver 108	(48) Cd Cadmium 112	(49) In Indium 115	(50) Sn Tin 119	(51) Sb Antimony 122	(52) Te Tellurium 128	(53) I Iodine 127	(54) Xe Xenon 131
6	(55) Cs Cesium 133	(56) Ba Barium 137	(57)-(71) RARE EARTH ELEMENTS	(72) Hf Hafnium 179	(73) Ta Tantalum 181	(74) W Tungsten 184	(75) Re Rhenium 186	(76) Os Osmium 190	(77) Ir Iridium 192	(78) Pt Platinum 195	(79) Au Gold 197	(80) Hg Mercury 201	(81) Tl Thallium 204	(82) Pb Lead 207	(83) Bi Bismuth 209	(84) Po Polonium 210	(85) At Astatine 210	(86) Rn Radon 222
7	(87) *Fr Francium 223	(88) Ra Radium 226	(89)-(103) ACTINIDE SERIES	(104) * Element 104 260	(105) * Element 105 260	(106) * Element 106 263	(107) * Element 107 261		(109) * Element 109 266									

RARE EARTH ELEMENTS	(57) La Lanthanum 139	(58) Ce Cerium 140	(59) Pr Praseodymium 141	(60) Nd Neodymium 144	(61) *Pm Promethium 147	(62) Sm Samarium 150	(63) Eu Europium 152	(64) Gd Gadolinum 157	(65) Tb Terbium 159	(66) Dy Dysprosium 163	(67) Ho Holmium 165	(68) Er Erbium 167	(69) Tm Thulium 169	(70) Yb Ytterbium 173	(71) Lu Lutetium 175
ACTINIDE SERIES	(89) Ac Actinium 227	(90) Th Thorium 232	(91) Pa Protectinium 231	(92) U Uranium 238	(93) *Np Neptunium 237	(94) *Pu Plutonium 242	(95) *Am Americium 243	(96) *Cm Curium 245	(97) *Bk Berkelium 245	(98) *Cf Californium 251	(99) *Es Einsteinium 254	(100) *Fm Fermium 253	(101) *Md Mendelevium 256	(102) *Mo Nobelium 254	(103) *Lw Lawrencium 257

Use the periodic table to answer questions 10 and 11.

10. Which of the following elements has the same number of shells as oxygen (O), atomic number 8? 10 ① ② ③ ④ ⑤

(1) Carbon (C)
(2) Sulfur (S)
(3) Helium (He)
(4) Aluminum (Al)
(5) Phosphorus (P)

11. Which of the following elements has the same number of electrons in its outermost shell as neon (Ne), atomic number 10? 11 ① ② ③ ④ ⑤

(1) Hydrogen (H)
(2) Oxygen (O)
(3) Helium (He)
(4) Nitrogen (N)
(5) Iodine (I)

For answers and explanations, see page 160.

COMPOUNDS AND MOLECULAR THEORY

KEY WORDS

element—a substance that cannot be broken into any simpler substances

compound—two or more elements that are combined chemically to form a single substance

bonding—the process by which a compound is formed

Forming Molecules

Very few of the elements on the periodic table exist in isolation. The vast majority of elements combine with other elements to form compounds.

As you have read, different elements have different numbers of electrons in their outer shells. Elements that have eight electrons in this outer shell have a stable structure and are called *inert*; they do not form compounds. Other elements do not have eight electrons in their outer shells; they are called *unstable*.

Compounds are formed because these unstable elements "try" to obtain a structure like the more stable inert elements. In effect, they are "looking around" for a place to gain or lose electrons. An atom that is "looking around" for an electron interacts with other atoms around it in order to get one. This interaction often results in atoms' bonding together. Bonding, a joining together of atoms, can be accomplished in one of two ways.

Bonding: Ionic and Covalent

The first method of forming a compound is called **ionic bonding** and can be explained by looking at the elements that make up ordinary table salt. These elements are sodium (Na) and chlorine (Cl). As you can see below, sodium has one electron in its outer shell and chlorine has seven.

SODIUM AND CHLORINE

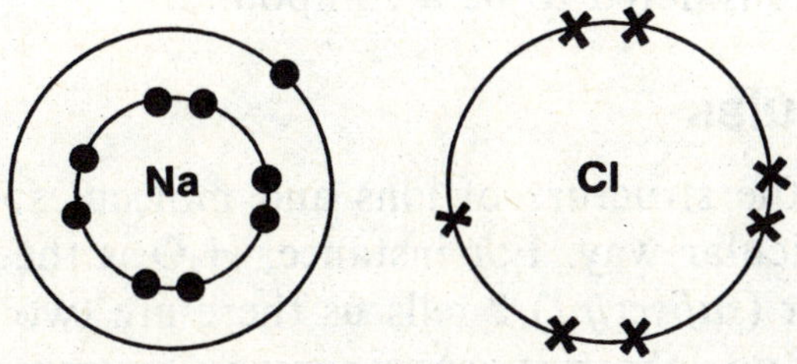

Because sodium "gives over" this electron, both the atom of sodium and the atom of chlorine obtain the stability they are "looking for." Each now has eight electrons in its outer shell. However, these atoms do not go their separate ways after this exchange. Remember that an electron has a negative charge. With this extra electron, chlorine now becomes negatively charged. Having gotten rid of one electron, the sodium atom has more protons than electrons and is therefore positively charged. Since opposites attract, the negative chlorine atoms bond to the positive sodium atoms, and sodium chloride, or salt, is formed.

SODIUM CHLORIDE

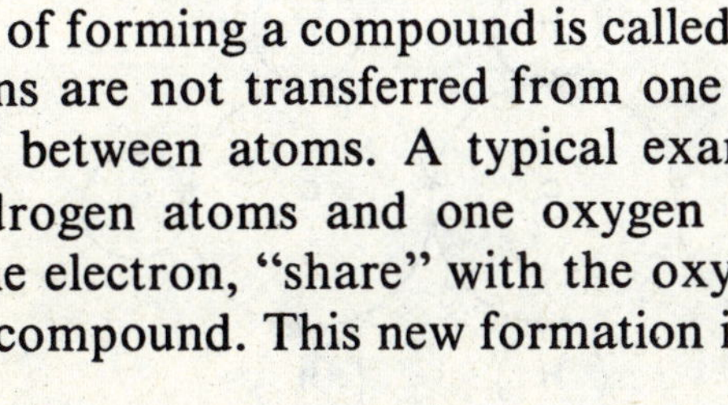

The other process of forming a compound is called **covalent bonding.** In this process, electrons are not transferred from one atom to another but rather are "shared" between atoms. A typical example is water. Water consists of two hydrogen atoms and one oxygen atom. The hydrogen atoms, each with one electron, "share" with the oxygen atom in order to make a more stable compound. This new formation is called a molecule of water.

A WATER MOLECULE

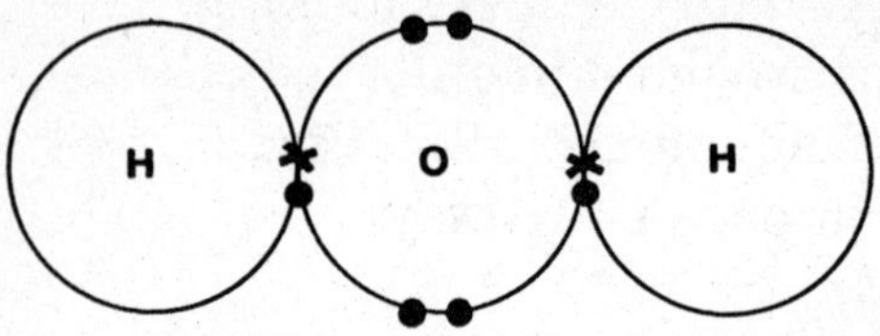

X = electrons from hydrogen
● = electrons from oxygen

The word molecule has two important meanings. One meaning is the joining of two or more atoms of different elements to form a compound, like the water molecule above. The other meaning is the joining of two or more atoms of the same element. This kind of molecule occurs when the atoms of one element tend to stick together in a certain number. A good example is the element hydrogen, which usually occurs as individual molecules of two atoms each. Because a hydrogen molecule consists of atoms of only one element, it is not considered to be a compound.

Formulas

Chemists use formulas to represent the structure of ions and molecules. These formulas are written in a particular way. For instance, H_2O is the formula for water. The small number (*subscript*) *2* tells us there are two atoms of hydrogen (H). Since there is no subscript accompanying oxygen, we know there is only one atom of oxygen (O). The formula for sodium chloride, NaCl, indicates that there is one atom of sodium (Na) and one of chlorine (Cl) in each molecule of salt. The formula for a molecule of hydrogen, which consists of two hydrogen atoms, is simply H_2.

In chemical equations, you may see a number in front of a molecule. If you see H_2O, you can assume that there is only one molecule of water. If the formula were written $4H_2O$, you would know that four molecules of water are indicated. If one molecule of water (H_2O) has two atoms of hydrogen and one atom of oxygen, then four molecules of water must have eight atoms of hydrogen ($4 \times 2 = 8$) and four atoms of oxygen ($4 \times 1 = 4$).

With molecules such as water, the structure of the molecule is fairly simple. For the more complex molecules, a diagram of the structure indicates what kinds of atoms are present and how they are connected. Propane, C_3H_8, is a fuel used for heating and in torches. The formula for propane tells you that it has three atoms of carbon and eight atoms of hydrogen, but it tells you nothing about how the atoms are arranged. The structure for C_3H_8 is shown below.

A MOLECULE OF PROPANE

X = electrons from hydrogen
● = electrons from oxygen

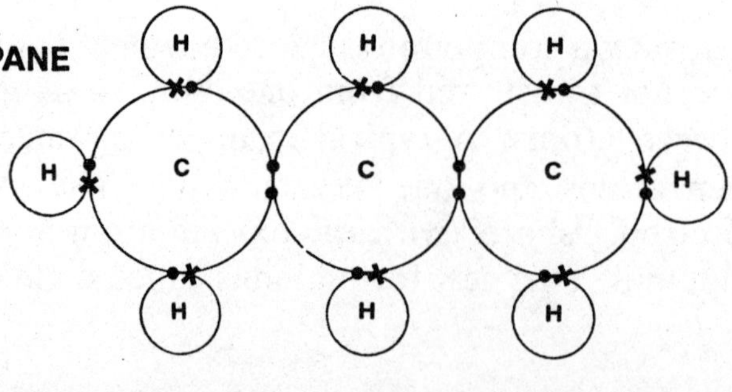

Directions: Based upon what you have just read, fill in the circle that corresponds to the correct answer.

12. An element with eight electrons in its outer shell is called
(1) sodium chloride
(2) inert
(3) negatively charged
(4) unstable
(5) a molecule

12 ① ② ③ ④ ⑤

13. The formula for carbonic acid is H_2CO_3. From this, you can tell that a molecule of carbonic acid has
(1) one atom each of hydrogen, carbon, and oxygen
(2) two molecules of hydrogen and two atoms of carbon
(3) two atoms of hydrogen, one of carbon, and three of oxygen
(4) a molecule of carbon dioxide and a molecule of water
(5) three atoms of hydrogen, two of carbon, and one of oxygen

13 ① ② ③ ④ ⑤

14. A covalent bond occurs when two elements
(1) share electrons
(2) give off electrons
(3) have opposite charges
(4) have more than eight electrons in their outer shells
(5) are inert

14 ① ② ③ ④ ⑤

15. Carbon dioxide consists of one atom of carbon (C) and two atoms of oxygen (O). The formula for a molecule of carbon dioxide is
(1) C_2O
(2) 2CO
(3) C_2O_2
(4) CO
(5) CO_2

15 ① ② ③ ④ ⑤

Use the following diagram to answer question 16.

CARBON TETRACHLORIDE

X = electrons from chlorine
● = electrons from carbon

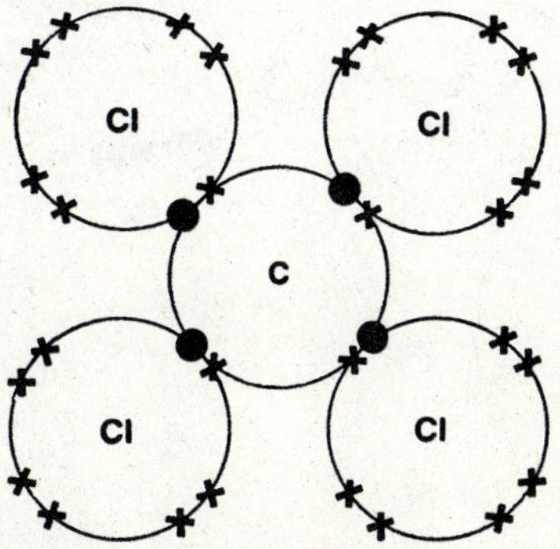

16. The diagram represents the structure of carbon tetrachloride, an industrial cleaner. Which of the following is the formula for one molecule of carbon tetrachloride?

16 ① ② ③ ④ ⑤

(1) C_4Cl
(2) $1C_4Cl$
(3) 4CCl
(4) CCl_4
(5) $4CCl_4$

For answers and explanations, see page 160.

CHEMISTRY EXERCISE 3

Directions: Read the passage and examine the diagram below. Answer the questions that follow. Fill in the circle that corresponds to the correct answer.

Molecular Formulas and Structure

Glucose is a sugar that is easily digested by humans. It is also the sugar that is manufactured by plants. The formula for glucose is $C_6H_{12}O_6$. The structure of glucose is shown in the diagram below. The diagram shows bonds with simple lines.

GLUCOSE

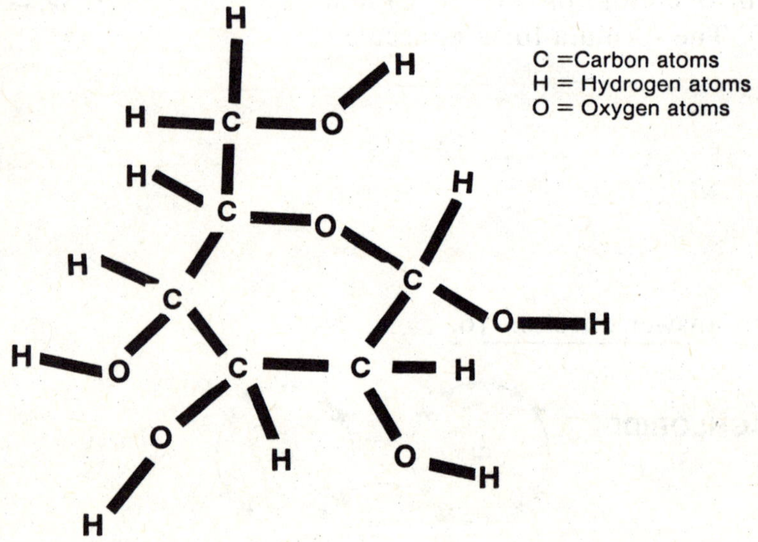

17. How many hydrogen atoms "share" their electrons with oxygen atoms?
(1) zero
(2) five
(3) six
(4) twelve
(5) twenty-four

17 ① ② ③ ④ ⑤

18. What is the total number of atoms necessary to make one molecule of glucose?
(1) three
(2) six
(3) twelve
(4) eighteen
(5) twenty-four

18 ① ② ③ ④ ⑤

19. What elements other than hydrogen are present in glucose?
(1) Oxygen and carbon
(2) Oxygen
(3) Carbon dioxide
(4) Carbon and water
(5) Glucose

19 ① ② ③ ④ ⑤

For answers and explanations, see page 161.

CHEMICAL REACTIONS AND THE CONSERVATION OF MATTER

KEY WORDS

equation—chemical equation that shows the elements that combine in a chemical reaction and the products that are formed

Law of Conservation of Matter—a fundamental scientific principle stating that matter cannot be created or destroyed in a chemical reaction

Chemical Reactions

When elements combine to form compounds, a chemical reaction takes place. The chemical reaction involves the breaking down of bonds between atoms and the formation of new bonds. A chemist indicates such a reaction with an equation:

$$2H_2 + O_2 \longrightarrow 2H_2O$$

This equation means that two molecules of hydrogen combine with one molecule of oxygen to yield two molecules of water.

The left side of the equation has two molecules of hydrogen (H_2) or a total of four hydrogen atoms. There is also one molecule of oxygen (O_2) or two atoms of oxygen. Chemical equations must be balanced, so the right side of the equation must have the same number of atoms as the left side. On the right side there are four hydrogen atoms and two atoms of oxygen.

Another example of a chemical reaction is photosynthesis. Photosynthesis is the process green plants use to manufacture food. Plants use carbon dioxide (CO_2), water (H_2O), and energy from the sun to create oxygen (O_2), glucose ($C_6H_{12}O_2$), and water.

$$\underset{\text{carbon dioxide}}{6CO_2} + \underset{\text{water}}{12H_2O} \xrightarrow{\text{light}} \underset{\text{oxygen}}{6O_2} + \underset{\text{glucose}}{C_6H_{12}O_6} + \underset{\text{water}}{6H_2O}$$

The 6 in front of the CO_2 indicates six molecules of carbon dioxide. The 12 in front of the H_2O shows twelve molecules of water.

The Law of Conservation of Matter

You will notice that the same elements are present on both sides of the equation. If there are six carbons on one side of the equation, there must be six on the other. There are twenty-four atoms of oxygen on each side.

If a chemical equation is to be valid, the actual number of atoms of each element must remain the same on both sides of the formula. Chemists call this the **Law of Conservation of Matter.** The law states that matter is neither created nor destroyed during chemical reactions. This law is a fundamental concept in chemistry.

We know that nitrogen combined with hydrogen (H) will give us ammonia (NH_3). Look at the molecules in this chemical reaction:

$$N_2 + 3H_2 \longrightarrow 2NH_3$$

Two atoms of nitrogen plus six atoms of hydrogen yield two molecules of ammonia. If you count the total number of atoms on each side, you will find them to be equal. This is an example of the Law of Conservation of Matter.

Directions: Based on what you have just read, fill in the circle that corresponds to the correct answer.

20. The Law of Conservation of Matter states that **20** ① ② ③ ④ ⑤
(1) matter is never changed chemically
(2) an unbalanced equation is an example of chemical change

(3) oxygen is necessary for all chemical reactions
(4) matter can be changed only when an acid is present
(5) matter is neither created nor destroyed during chemical reactions

21. The equation $C + O_2 \longrightarrow CO_2$ 21 ① ② ③ ④ ⑤
(1) is an example of an unbalanced equation
(2) directly contradicts the Law of Conservation of Matter
(3) means that carbon and oxygen create two new atoms
(4) indicates three atoms on both the left and the right sides of the equation
(5) has one carbon atom on the left but two on the right

22. In the formula for photosynthesis, how many molecules of oxygen (O_2) are released? 22 ① ② ③ ④ ⑤

$$6CO_2 + 12H_2O \longrightarrow 6O_2 + C_6H_{12}O_6 + 6H_2O$$

(1) One
(2) Two
(3) Six
(4) Twelve
(5) None

For answers and explanations, see page 161.

CHEMISTRY EXERCISE 4

Directions: Read the passage below and answer the questions that follow. Fill in the circle that corresponds to the correct answer.

Exhaust Pollutant "Neutralized"

Ellagic acid—a compound that belongs to a naturally occurring chemical class found in a variety of sources, including coffee, nuts, and grapes—has been shown to "neutralize" the cancer-causing form of the environmental pollutant benzo(a)pyrene. This phenomenon has thus far been confirmed only in laboratory tests, scientists say. Nonetheless, it is hoped that further research on ellagic acid's apparent ability to conquer the dangerous form of benzo(a)pyrene may lead to a new class of cancer-preventing drugs.

Benzo(a)pyrene is a ubiquitous pollutant formed from the incomplete burning of fossil fuels. The chemical is found, for example, in automotive exhaust and industrial smokestack emissions. When the chemical enters the body, enzymes can convert it to a potent mutagen and carcinogen.

Chemists have found that, when ellagic acid is added to a solution of this potent form of benzo(a)pyrene, the two compounds first join to form an inactive complex. Eventually, the addition of ellagic acid causes the carcinogen to split into noncancer-causing, nonmutagenic components. Scientists are now conducting experiments to determine whether ellagic acid protects animals from getting tumors after exposure to benzo(a)pyrene.

23. Ellagic acid is able to "neutralize" the pollutant called benzo(a)pyrene

(1) by destroying it
(2) by combining with it to form nonmutagenic components
(3) by converting automotive exhaust into oxygen
(4) only if the pollutant exists in the atmosphere
(5) only if the two are burned together

23 ① ② ③ ④ ⑤

24. Which of the following is most likely to be a source of ellagic acid?

(1) Almonds
(2) Milk
(3) Fish
(4) Tea
(5) Bread

24 ① ② ③ ④ ⑤

25. After chemical tests confirm that a drug may prevent cancer, what is the next step that scientists take to prove that a drug can prevent cancer?

(1) Experiment on animals
(2) Experiment on humans
(3) Write chemical formulas
(4) Pass a law that permits use of the substance
(5) No more steps are necessary

25 ① ② ③ ④ ⑤

For answers and explanations, see page 161.

ORGANIC CHEMISTRY

KEY WORDS

hydrocarbons—compounds made up of hydrogen and carbon. Fuels are forms of hydrocarbons.

organic compound—any compound containing carbon

The Carbon Cycle

Carbon, an element found in abundance on earth, is so important to us that chemists have named the study of carbon *organic chemistry*, sometimes called "the chemistry of life." Carbon atoms can easily bond covalently with other carbon atoms. These carbon molecules can then bond with more carbon atoms, forming even longer chains of carbon molecules. Carbon also forms strong bonds with other vital elements, such as hydrogen, oxygen, and nitrogen.

The carbon cycle is very important to life. Plants and animals are made up of large amounts of carbon. When these organisms die, they decompose. The carbon atoms can do one of two things. Some combine with the oxygen in the atmosphere to form carbon dioxide, the key element in photosynthesis. Other carbon atoms become trapped under the earth, cut off from oxygen. Over a period of centuries, large deposits of carbon are transformed into fossil fuels such as petroleum, natural gas, and coal.

Petroleum and coal that are removed from the earth can be burned or distilled to meet human needs. Coal is heated in the absence of air, leaving coke, which is used to separate metals from ores. Coal tar, which is also produced in this heating process, is an important base of many **organic compounds.**

When petroleum is purified through distillation, **hydrocarbons** (compounds of hydrogen and carbon) separate to form different types of fuel. Methane (CH_4) and butane (C_4H_{10}) are important in the production of plastics. Kerosene ($C_{12}H_{26}$) is used in diesel fuel and home heating fuel. The distillation of petroleum produces the hydrocarbon compounds that make up perfumes, pesticides, preservatives, medicines, and many other items we consider vital.

By-products of Carbon Compounds

When the hydrocarbons in gasoline are burned, such as in the engine of an automobile, they do not disappear but instead combine with oxygen in the air to form carbon dioxide and water. We know that these compounds are safe. However, a danger arises when there is not sufficient oxygen with which carbon can combine. Instead of forming carbon dioxide, these hydrocarbons form carbon monoxide, a poisonous gas. Carbon monoxide is a common form of air pollution produced by automobiles and is particularly dangerous where traffic is heavy.

If you take a look at the trash that pollutes our land, you will find that the majority of this litter is made up of plastics. We have ingeniously created compounds that can hold up under extreme heat, pressure, and moisture. We have made plastic bags, cellophane wrapping, six-pack holders, automobile tires, foam rubber, and countless other products. The problem is, what do we do with these materials when they are no longer useful? We can throw them away, but solid waste dumps are unsanitary and use up valuable land. We can burn them, but burning plastic creates thick black smoke that is another dangerous form of air pollution.

It seems as though the Law of Conservation of Matter has caught up with us. We cannot destroy matter. As we consume our natural resources, such as petroleum and coal, we create useful products, but we also create unwanted by-products that do not just disappear. We are now realizing that it is our responsibility to find ways to dispose of these unwanted materials. It is clear that organic chemists must now also play a role in discovering safe ways to deal with the dangerous by-products of their work.

Directions: Based upon what you have just read, fill in the circle that corresponds to the correct answer.

26. Which of the following is a true statement about organic chemistry?

(1) Organic chemistry is relatively insignificant when compared to other fields of chemistry.
(2) Organic chemistry is the study of carbon and its compounds.
(3) The study of organic chemistry is concerned mainly with pollution.
(4) The field of organic chemistry got its name because it is the study of human internal organs.
(5) Organic chemistry is the only significant area of study in the entire field of chemistry.

26 ① ② ③ ④ ⑤

27. From the essay, you can infer that

(1) organic chemists can play a key role in solving our pollution problems
(2) pollution is caused by our inability to get rid of hydrocarbons
(3) the by-products of our chemical advances and discoveries are always beneficial to humans
(4) carbon monoxide poisoning is less of a problem in the East than in the Midwest
(5) plastic refuse will not be a problem in the future because the carbon atoms will decompose into the ground and create new petroleum

27 ① ② ③ ④ ⑤

28. Coal

28 ① ② ③ ④ ⑤

(1) must undergo complicated processing before it can be used to heat homes
(2) can be used in place of natural gas to avoid many of the dangers of pollution
(3) is formed over a period of thousands of years when carbon combines with oxygen under the earth
(4) is heated in the absence of air to produce coke, which is used to refine metal
(5) is necessary in the production of plastics

For answers and explanations, see page 161.

CHEMISTRY EXERCISE 5

Directions: Read the passage below and answer the questions that follow. Fill in the circle that corresponds to the correct answer.

Alcohol Alchemy

Hold on to your vodka martini—the alcohol in it, called *ethanol*, can now be converted easily into gasoline.

Although pure alcohol can be used as fuel in specially modified engines, it does not provide as much energy as an equivalent weight of gasoline. Unlike gas, which consists entirely of carbon and hydrocarbon atoms, alcohol also contains oxygen, which lowers its energy level.

Scientists have devised an effective way to take the oxygen out of ethanol and use the remaining atoms to make gasoline.

According to the scientists working on the project, the process involves heating ethanol and water and passing the resulting vapor through a hot bed of a mineral called zeolite. In a complex chemical reaction catalyzed by the zeolite, oxygen in the ethanol combines with hydrogen to form water. The carbon and any remaining hydrogen join to form gasoline. The gas and water then collect in a tank and are separated. Interestingly, a mixture of water and ethanol is exactly what is produced in the process of fermenting grain. Scientists say that someday farmers could make their own gasoline by hooking up an ethanol-to-gas converter to a modern version of the moonshiner's still.

29. According to the passage, gasoline can be made from alcohol

29 ① ② ③ ④ ⑤

(1) by burning a mixture of alcohol and water
(2) if vodka is used for experimentation and not for drinking
(3) but only for use in farm machinery and other modified engines
(4) by taking oxygen out of ethanol, leaving behind the necessary hydrogen and carbon
(5) without any chemical changes taking place

30. Why is pure alcohol unsuitable as a fuel?

30 ① ② ③ ④ ⑤

(1) It is much too expensive. Making gas from it would not be cost effective.
(2) It contains too many carbons and not enough hydrogen.
(3) The zeolite it contains is damaging to the metals in an engine.
(4) There is simply not enough of it for both drinking and making gas.
(5) Alcohol contains oxygen, which lowers its energy level.

For answers and explanations, see page 162.

SOLUTIONS

KEY WORDS

electrolyte—a substance that conducts electric current when in liquid form

alloy—a material composed of two or more metals

What Is a Solution?

What happens when salt is placed in water? Although the salt seems to have disappeared, the Law of Conservation of Matter tells us that it can't. The salt has been broken up, or dissolved, by the water. The resulting salty water is called a *solution*. Sand in water would not be considered a solution because the sand does not dissolve. After a short time, sand in water simply settles to the bottom.

Solutions have practical applications. Every soap or detergent product you use relies on solutions. Water alone cannot dissolve grease or dirt, but soap molecules dissolve in both water and grease. Soap is used to "hook together" water and grease molecules, which do not normally mix. When grease molecules are linked to water, they can be washed away.

Electrolytes

When substances with ionic bonding are dissolved, they are broken down into ions. An ion is an atom that has either a positive or a negative charge. When salt (NaCl) dissolves, it forms positive ions ($Na+$) and negative ions ($Cl-$). The NaCl solution conducts electric current. A substance that conducts an electric current when dissolved in water (such as salt) is called an **electrolyte**.

Electrolytes can be divided into three types of substances: acids, bases, and salts. We know NaCl is a salt. Let's look at the characteristics of the other two substances. An acid in a water solution has a sour taste, like lemon juice or vinegar. An example of this is HCl (hydrochloric acid), the acid that helps break down food in your stomach. A base has a very soapy or slippery feel to it. Detergents and lye are examples of bases. When an acid is added to a base, the two react and form a salt and water.

Nonliquid Solutions

Although we usually think of solutions as containing a solid dissolved in a liquid, solutions involving other states of matter also exist. Carbon dioxide is pumped under pressure into flavored water to produce carbonated beverages, an example of a gas dissolved in a liquid.

Metal **alloys** are solid solutions that are made by melting different metals together. This process produces a new metal that has the desirable characteristics of the metals used. Brass is made of copper and zinc. Copper is best known for its ability to conduct electricity. It is also very easy to shape. The addition of zinc makes the resulting brass stronger and tougher. The dental amalgam used to fill cavities contains silver and mercury. Silver does not corrode or decay but is difficult to work. Mercury is added to allow the amalgam to be shaped.

Directions: Based on what you have just read, fill in the circle that corresponds to the correct answer.

31. When a substance is dissolved in another substance, the result is a(n)

(1) base
(2) acid
(3) molecule
(4) solution
(5) salt

31 ① ② ③ ④ ⑤

32. According to the passage, when substances with ionic bonds are dissolved in water, the solution can conduct electric current. Which of the following terms is used to identify these substances?
(1) Molecules
(2) Electrolytes
(3) Hydrocarbons
(4) Neutralizers
(5) Organic compounds

32 ① ② ③ ④ ⑤

CHEMISTRY EXERCISE 6

Directions: Read the passage below and answer the questions that follow. Fill in the circle that corresponds to the correct answer.

Dry Cleaning

There is nothing "dry" about dry cleaning. This process was probably called "dry" because it does not involve water, as the more common "wet" laundry technique does. Why we need these different methods to clean our clothes can easily be explained by the chemical reactions that go on in both a dry cleaning plant and a washing machine.

Soap molecules "hitch up" water molecules with the grease and dirt in your clothes and wash the dirt away. The fibers of the fabric must swell up in order to absorb the soap and water necessary for cleansing. As the fabric dries, some of the fibers shrink down to their original shape and size, but others do not, causing the shrinkage and wrinkling we notice after our clothing has been laundered. Even though water is sometimes called the "universal solvent," it does have some drawbacks.

Dry cleaners use a different solvent to combat this problem. It is called tetrachloroethylene (C_2Cl_4). This solvent is a very powerful dissolver of grease and other staining substances. It has an important advantage over soap and water that keeps the dry cleaner in business. C_2Cl_4 evaporates faster than water—so quickly, in fact, that you can actually watch a drop of this solvent disappear in a matter of minutes. The carbon and chloride atoms combine quickly with warm air, leaving the fabric dry in a very short period of time with none of the shrinkage and wrinkles that accompany water solvents.

33 ① ② ③ ④ ⑤

33. According to this passage, which of the following is the advantage of dry cleaning over washing with water?
(1) Soap does not build up in the fabric.
(2) Clothes are much less likely to shrink or wrinkle.
(3) Tetrachloroethylene is not effective on tough stains.
(4) Dry cleaning is less expensive because water is not needed.
(5) Soap and water wear out the fabric.

34 ① ② ③ ④ ⑤

34. What advantage does tetrachloroethylene (C_2Cl_4) have over soap and water?
(1) C_2Cl_4 evaporates more quickly than water.
(2) C_2Cl_4 causes fibers to swell up, while water does not.
(3) C_2Cl_4 is inexpensive compared to detergent.
(4) C_2Cl_4 causes the fabric to shrink, while water does not.
(5) C_2Cl_4 takes longer to evaporate than water.

For answers and explanations, see page 162.

ANSWERS AND EXPLANATIONS

WHAT IS CHEMISTRY?

Making Inferences **1.** (5) The second paragraph discusses the tremendous impact chemistry has had on the development and growth of the plastics industry. You can infer that chemistry was involved in the development of transparent plastics. The field of chemistry is broader than just the manufacture of drugs, choice (1).

Finding Details **2.** (3) The first paragraph states that, in order to study chemistry, matter must be broken down into its simplest form.

Finding Details **3.** (4) The last paragraph describes the development of aspirin from a natural treatment that used the bark of the willow tree to a man-made chemical compound called *acetylsalicylic acid.*

CHEMISTRY EXERCISE 1

Main Idea **4.** (5) The four steps in the scientific method are listed in the reading passage.

Making Inferences **5.** (3) From the first three paragraphs, you can infer that the alchemists were not interested in developing accurate methods but were more interested in the possible outcomes of their efforts—creating gold and curing all illnesses.

ATOMS AND ATOMIC THEORY

Finding Details **6.** (2) The answer is stated specifically in the first paragraph of this essay.

Finding Details **7.** (1) The last paragraph states that the atomic number is equal to the number of protons in an atom.

Applying Ideas **8.** (3) The atomic number is equal to the number of protons. If helium has two protons, then the atomic number is two.

CHEMISTRY EXERCISE 2

Finding Details **9.** (5) The second paragraph discusses how the elements are arranged on the periodic table—according to the number of electrons and the number of shells.

Applying Ideas **10.** (1) Lines 16–17 state that elements in each period, or horizontal row, have the same number of shells. Of the elements listed, only carbon is on the same horizontal row as oxygen.

Applying Ideas **11.** (3) Lines 15–16 state that elements in each group, or vertical column, have the same number of electrons in their outermost shell. Of the elements listed, only helium is in the same vertical column as neon.

COMPOUNDS AND MOLECULAR THEORY

Finding Details **12.** (2) The second paragraph of the essay refers to elements with eight electrons in their outermost shells as *inert* or *stable.*

Applying Ideas **13.** (3) Carbonic acid (H_2CO_3) has two atoms of hydrogen (H_2), one atom of carbon (C), and three atoms of oxygen (O_3).

Finding Details **14.** (1) The sixth paragraph states that in covalent bonding electrons do not transfer, choice (2), but rather are shared.

Applying Ideas **15.** (5) One atom of carbon would be written *C* and two atoms of oxygen would be written O_2. Thus, the formula for carbon dioxide is CO_2.

Applying Ideas **16.** (4) The diagram of carbon tetrachloride shows one atom of carbon (C) and four atoms of chlorine (Cl_4). The formula would be written CCl_4.

CHEMISTRY EXERCISE 3

Applying Ideas **17.** (2) Only five of the hydrogen atoms "share" their electrons with oxygen. The other seven "share" electrons with carbon.

Finding Details **18.** (5) There are six carbon atoms (C_6), twelve hydrogen atoms (H_{12}), and six oxygen atoms (O_6). This makes a total of twenty-four atoms.

Finding Details **19.** (1) Glucose has only three elements: hydrogen, oxygen, and carbon. Choice (2) is incomplete, and the other choices are compounds, not elements.

CHEMICAL REACTIONS AND THE CONSERVATION OF MATTER

Main Idea **20.** (5) The Law of Conservation of Matter is discussed in the fifth paragraph.

Applying Ideas **21.** (4) The left side of the equation shows one atom of carbon and two atoms of carbon. The right side of the equation also has three atoms. Since the equation is balanced, choices (1) and (2) are not true.

Applying Ideas **22.** (3) Look for oxygen (O_2) on the right side of the equation. The "$6O_2$" indicates that six molecules of oxygen are released.

CHEMISTRY EXERCISE 4

Finding Details **23.** (2) Lines 15–19 state that, when ellagic acid is added to benzo(a)pyrene, an inactive nonmutagenic complex is formed. The pollutant is not destroyed, choice (1), only "neutralized," or made inactive.

Drawing Conclusions **24.** (1) The first paragraph mentions several food products that contain ellagic acid, including nuts. Almonds are nuts.

Making Inferences **25.** (1) The passage mentions the laboratory experiments that were conducted on ellagic acid. Lines 19–22 mention that scientists are now conducting further experiments using animals. From this information, you can infer that the testing of a substance on animals is the next step in the process of proving that a drug can prevent cancer.

ORGANIC CHEMISTRY

Finding Details **26.** (2) The first paragraph of this essay states that organic chemistry is the study of carbon and compounds that are formed with carbon.

Making Inferences **27.** (1) The final paragraph gives you the basis to infer that organic chemists can play a key role in resolving pollution problems. We have no difficulty getting rid of hydrocarbons, choice (2); it is the by-products of hydrocarbons that cause difficulties.

Finding Details **28.** (4) The third paragraph discusses the uses of coal.

CHEMISTRY EXERCISE 5

Finding Details **29.** (4) Lines 8–9 state that scientists have devised an effective way to take the oxygen out of ethanol and use the remaining atoms to make gasoline.

Drawing Conclusions **30.** (5) Lines 6–7 state that alcohol does not provide as much energy as an equivalent weight of gasoline. The difference is accounted for by the presence of oxygen atoms in alcohol.

SOLUTIONS

Main Idea **31.** (4) The passage discusses various types of solutions. The first paragraph discusses a solution with salt dissolved in water.

Finding Details **32.** (2) According to the section on electrolytes, an electrolyte is a substance that conducts an electric current when dissolved in water.

CHEMISTRY EXERCISE 6

Drawing Conclusions **33.** (2) While choices (1), (3), and (5) may be true, they are not mentioned in the passage. The last line mentions that C_2Cl_4 does not have the problem of shrinkage and wrinkling.

Finding Details **34.** (1) The third paragraph states that C_2Cl_4 has an important advantage over soap and water in that it evaporates faster than water.

Physics

INTRODUCTION

Physics is the study of matter, energy, force, space, and time. This science explains how our world is put together and how it changes. The field of physics overlaps other sciences, such as geology, astronomy, and chemistry. For example, an astrophysicist studies the location and physical properties of stars, planets, and other objects in the sky. A geophysicist studies the movement of the earth's crust, the occurrence of earthquakes, and the formation of volcanoes.

The physics portion of the GED science test covers mechanics, matter and heat, optics and waves, and electricity and magnetism. Test questions will center around the practical application of physics concepts. You do not need to know algebra or geometry to answer these questions.

The science of motion and the application of forces to material objects is called **mechanics**. The principles of mechanics guide engineers when they design structures such as high-rise buildings, highways, and spacecraft.

The study of **matter and heat** involves not only the structures and properties of matter but also how matter responds to heat. For instance, heat may cause matter to change form, such as water turning to steam when it is heated.

The studies of light and sound are related through **wave theory** because both light and sound travel as waves. The study of waves has led to significant advances in science. For example, scientists have discovered ultrasound, sound waves that are too high in frequency to be heard by humans. These waves can be bounced off the surface of an object and used to detect the object's shape. One important use of ultrasound is in medicine: doctors use it to detect tumors in patients and to monitor the growth of fetuses in pregnant women. Important advances resulting from wave theory have also been made in the field of **optics**, the study of the behavior of light. These advances have led to the invention of modern microscopes, telescopes, and cameras.

Using what they know about the relationship of **electricity and magnetism**, scientists and engineers have been able to build generators to produce electricity, paving the way for many labor-saving devices. Electronics has taken us from the study of electrons in vacuum tubes to the development of microprocessors that contain thousands of electronic circuits on a small silicon chip. The development of microprocessors has been instrumental in the rapid growth of the computer industry.

Directions: Based on what you have just read, fill in the circle that corresponds to the correct answer.

1. While biology is the science of living things, physics studies
   (1) meteorology and oceanography
   (2) astronomy
   (3) matter, energy, force, space, and time
   (4) logic and mathematics
   (5) carbon compounds

1 ① ② ③ ④ ⑤

2. Scientists interested in finding out if a new plastic could serve as a replacement for glass in eyeglasses would be most interested in which area of physics?
   (1) Optics
   (2) Mechanics
   (3) Matter and heat
   (4) Magnetism
   (5) Electronics

2 ① ② ③ ④ ⑤

3. Which of the following areas of physics would be of primary concern to an engineer designing the body of a small airplane?
   (1) Matter and heat
   (2) Ultrasound
   (3) Microprocessors
   (4) Mechanics
   (5) Optics

3 ① ② ③ ④ ⑤

For answers and explanations, see page 181.

THE LAWS OF MOTION

KEY WORDS

friction—the resistance of one object rubbing against another

gravity—a natural force that pulls all objects toward the center of the earth

inertia—the tendency of an object either to remain motionless or to remain in motion at the same speed and in the same direction

mass—the amount of matter in a body

Sir Isaac Newton (1642–1727), whose work has greatly influenced modern physics, identified three laws that govern motion. Newton theorized that these laws operate best in a vacuum. A vacuum is a space from which all air has been removed. Objects in a vacuum do not experience air **friction**, the resistance met when a moving object collides with air molecules. Newton's laws of motion also apply in everyday life, but they are not as noticeable because of the influences of outside forces such as **gravity** and friction.

Newton's First Law

Newton's first law states that a body at rest remains at rest and that a body in motion continues to move in a straight line unless some outside force affects it. This property is called **inertia**. The magician's trick of pulling a tablecloth out from underneath a fully set table without disturbing the setting is an example of the principle of inertia.

Newton's Second Law

The second law of motion states that a change in the motion of an object depends upon the **mass** of that object and the amount of the force applied to it. You can separate this law into two parts: (1) An object's speed will increase in proportion to the amount of force applied to it. The stronger the force, the faster the object's speed will increase. (2) However, this acceleration will also be affected by the mass of the object. The larger the mass, the slower the object will accelerate. The smaller the mass, the faster the acceleration.

To understand the relationship between mass and force, imagine a man struggling to push an elephant. Then imagine the man using the same force to push a small dog. The elephant would barely move, while the dog could be moved easily.

Newton's Third Law

Newton's third law of motion says that for every action there is an equal and opposite reaction. The recoil or kick of a rifle illustrates this law. Gunpowder explodes in a rifle, forcing the bullet through the barrel. The exploding powder is the force that pushes the bullet through the barrel. It is also the force that pushes the rifle backward against the shooter's shoulder.

Directions: Based on what you have just read, fill in the circle that corresponds to the correct answer.

4. Newton's first law of motion accounts for the fact that — 4 ① ② ③ ④ ⑤
(1) a basketball rebounds off a backboard
(2) a car skidding on a wet curve stays on the road
(3) a kite stays in the air on a windy day
(4) a car without brakes is difficult to stop
(5) cats can climb up trees but can't climb down

5. From the discussion of Newton's second law of motion, you can infer that the word *acceleration* means — 5 ① ② ③ ④ ⑤
(1) an increase in speed
(2) either large or small
(3) motion
(4) mass
(5) a great force

6. Isaac Newton identified three laws of motion. In other words, — 6 ① ② ③ ④ ⑤
(1) the laws did not exist until he wrote them
(2) he invented the actions that are described by the laws
(3) he recognized and explained basic principles governing the motion of objects
(4) the laws had to be approved by a court of law
(5) he recognized that three basic types of motion existed and had them written and passed by the British Parliament

7. A pendulum would act differently outside of a vacuum because which of the following would be working on it? — 7 ① ② ③ ④ ⑤
(1) Water
(2) Acceleration
(3) Light
(4) Inertia
(5) Friction

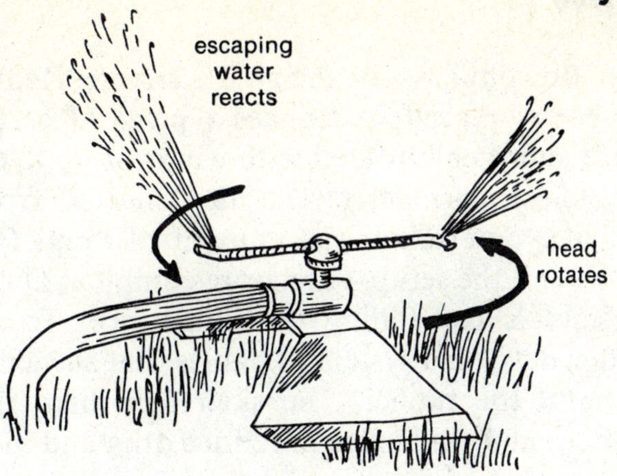

The reaction force of the water against the nozzles causes the sprinkler to rotate.

8. According to the diagram above, the action of the sprinkler illustrates

(1) Newton's first law of motion: inertia
(2) Newton's second law of motion: the change in motion of an object depends on the amount of force and the mass of that object
(3) Newton's third law of motion: for every action there is an equal and opposite reaction
(4) Newton's law of gravity
(5) that a body at rest stays at rest

8 ① ② ③ ④ ⑤

For answers and explanations, see page 181.

PHYSICS EXERCISE 1

Directions: Read the passage below and answer the questions that follow. Fill in the circle that corresponds to the correct answer.

Wind Tunnel Cuts Skiers' Drag

The ski jumper feels the rush of wind roaring past as he floats suspended in midair, his slightest move monitored on a video screen and meters embedded in the floor below. He and the rest of the U.S. Olympic Ski Team are streamlining their style in preparation for the 1984 games. And they're doing it with the help of a 120-foot-long wind tunnel at Calspan Advanced Technology Center in Buffalo, New York.

Michael Holden, principal aeronautical engineer at Calspan—and a skier himself—convinced the firm to let the team use one of their wind tunnels so they could pare crucial fractions of a second from their race times. "The tunnel simulates the conditions a skier faces," says Holden. "The most effective aerodynamic position is different in each situation."

For downhill racers, the goal is simple: Reduce drag. Each team member is placed in stances typical of certain sections of the course and then buffeted with winds of up to 60 miles per hour. A series of sensors detect the air deflected from his body, and a computer determines the amount of drag. For ski jumpers, the goals—and the setup—are more complex. During an actual jump, the skier is free-falling through space. To recreate this effect, Holden devised a system of cables that suspend the skier from the ceiling of the tunnel. The skier can then try out various body positions and ski angles to reduce drag and increase lift—in effect making his body into a gliding wing.

Holden got the idea while attending the 1980 Winter Olympics in Lake Placid, New York. "Prior to this training method, a skier learned by watching others. Now he learns by the feel of the air against his body and can respond to similar conditions on the hill."

9. The U.S. Olympic Ski Team uses a wind tunnel to prepare for the Olympics. What is the goal of the downhill racers?

(1) To increase friction
(2) To free-fall through space
(3) To simulate flying conditions
(4) To ski in a vacuum
(5) To reduce drag

9 ① ② ③ ④ ⑤

10. Why are the Olympic skiers using a wind tunnel to simulate ski conditions?

(1) To try to fly
(2) To watch themselves on video monitors
(3) To watch other skiers
(4) To perfect the wind tunnel
(5) To improve their skiing performance

10 ① ② ③ ④ ⑤

SKIER IN FLIGHT

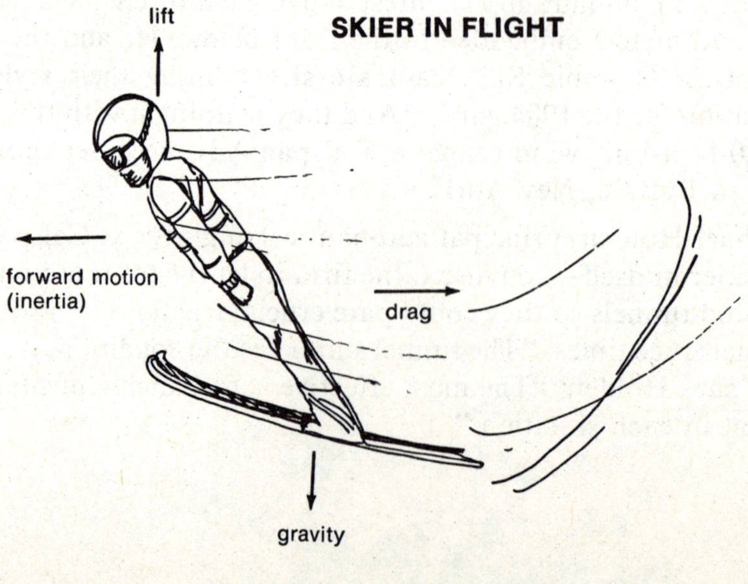

11 ① ② ③ ④ ⑤

11. The diagram on the preceding page shows the different forces on a ski jumper in the air. According to the diagram, what force opposes the forward motion of the ski jumper?
(1) Gravity
(2) Drag
(3) Lift
(4) Inertia
(5) Acceleration

For answers and explanations, see page 181.

STATES OF MATTER

KEY WORDS

diffusion—the spreading of the molecules of one substance throughout the molecules of another

condense—to change from a gas or vapor to a liquid

Matter is defined simply as anything that occupies space. It can exist in one of three forms or states at any one time: solid, liquid, or gas. Matter can change form, and each state of matter has its own special characteristics or properties. Different substances change from one state of matter to another at different temperatures.

The molecules of solid matter are strongly attracted to each other and are very close together. The volume of a solid is constant because it does not change shape easily. A piece of steel and a rock are both solids; each is a firm, unchanging object. If you place a solid in a container, it will go to the bottom of the container and will not change shape.

Although the attractive forces among the molecules of a liquid are strong enough to give it a definite volume, the molecules move more freely than in a solid, and the liquid can change shape depending on the container it is in. However, a liquid's volume and the amount of space it occupies do not change. If you knock over an eight-ounce glass of water, the water will look different lying in a puddle than it did in the glass, but the amount of water will not have changed. If you place the liquid in a container, it will go to the bottom and take the shape of the container.

The molecules of a gas move very freely, allowing for rapid changes in shape and volume. If a gas is sealed in a container, it takes on the shape of the container. Once the gas is released from the container, the molecules spread over a larger volume of space. This effect is called **diffusion**. Gas will diffuse until it uniformly fills the object that contains it. It is also true that a gas at a higher temperature will diffuse faster than a gas at lower temperature.

Solids change to liquids and liquids change to gases at certain temperatures. Ice melts at temperatures above 32° F. If you remove an ice cube from the freezer and leave it in a glass at room temperature, it will melt, becoming a liquid. If you pour the water into a pot on the stove and heat it to 212° F, the water will become a gas, known as *water vapor*. When the air cools, the vapor **condenses** and becomes a liquid again. The water may be frozen again to form ice, a solid.

Directions: Based on what you have just read, fill in the circle that corresponds to the correct answer.

12. The three states of matter are
(1) ice, water, and water vapor
(2) fire, water, and temperatures
(3) solid, liquid, and molecules
(4) attraction, force, and pressure
(5) solid, liquid, and gas

12 ① ② ③ ④ ⑤

13. Granules of instant coffee are an example of which state of matter?
(1) A solid
(2) A liquid
(3) An evaporating liquid
(4) A gas
(5) A gas that is condensing

13 ① ② ③ ④ ⑤

14. If a gas is confined in a container, the
(1) pressure begins to decrease
(2) gas takes on the shape of the container
(3) gas becomes a liquid
(4) temperature begins to rise
(5) attraction between the molecules is stronger than in a solid

14 ① ② ③ ④ ⑤

For answers and explanations, see page 181.

PHYSICS EXERCISE 2

Directions: Read the passage below, examine the accompanying diagram, and answer the questions that follow. Fill in the circle that corresponds to the correct answer.

Refrigerators

The most baffling household appliance is the refrigerator. The workings of other household appliances can usually be understood, but the refrigerator, with its seemingly endless tubing, is hard to comprehend by just looking at its parts.

A refrigerator relies on the compression of gases and the process of evaporation. When a gas is compressed, its temperature rises. When a gas is allowed to expand, or when a liquid evaporates, heat is absorbed by the gas.

The compressor pumps a gas, usually freon, through the entire system. The compressor compresses the freon and pumps it into the discharge line (number **1** on the diagram). While the gas is being compressed, its temperature rises. The heated gas enters the condenser (**2**), usually on the back or bottom of the refrigerator, where much of the heat is released into the air. As the compressed gas releases its heat, it cools and condenses into a liquid.

A narrow tube (**3**) carries the liquid freon from the condenser (**2**) to the evaporator tubing (**4**). The evaporator tubing is larger than the tube from the condenser so that as the liquid enters the evaporator tubing it begins to expand and becomes a gas. In the process of passing through the evaporator tubing, the evaporating liquid absorbs the heat inside the refrigerator and cools the contents of the refrigerator.

The gas returns to the compressor through the suction line (**5**), to begin another refrigeration cycle.

THE COMPRESSION SYSTEM OF A REFRIGERATOR

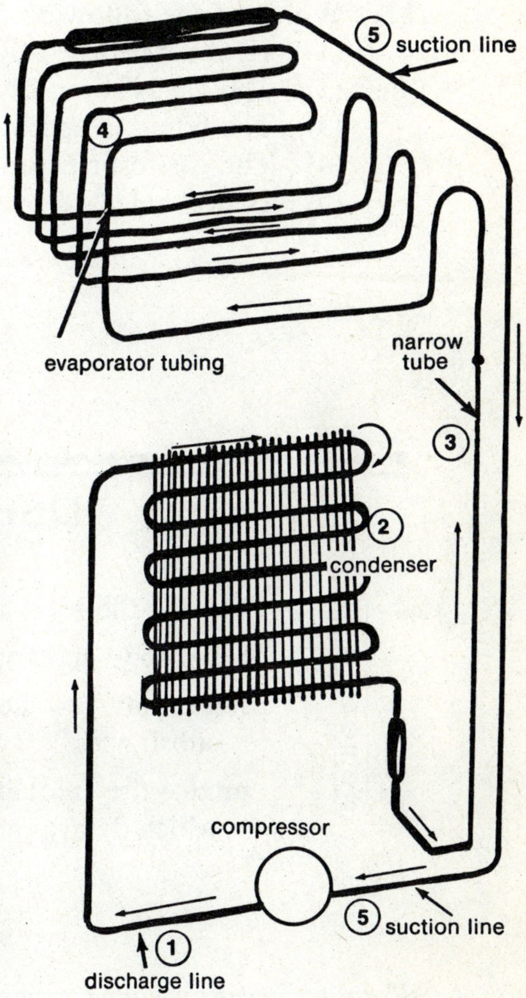

15. Which of the following gases is most frequently used in home refrigerators?
(1) Freon
(2) Argon
(3) Ammonia
(4) Hydrogen
(5) Helium

15 ① ② ③ ④ ⑤

16. In a closed system such as a refrigerator, a liquid that is allowed to evaporate will

16 ① ② ③ ④ ⑤

(1) be compressed
(2) remain a liquid
(3) absorb heat
(4) release heat
(5) be released into the air

17. Which of the following occurs in the condenser?

17 ① ② ③ ④ ⑤

(1) Heat is absorbed by the gas.
(2) The gas is compressed.
(3) The liquid evaporates into a gas.
(4) The gas returns to the compressor.
(5) Heat is released by the compressed gas.

For answers and explanations, see page 182.

LIGHT AND SOUND

KEY WORDS

waves—the form in which light and sound travel

reflection—light striking a surface and changing direction

refraction—the bending of light as it passes through different substances

pitch—the frequency of vibrations, or waves, of any sound, which determines how high or low the sound is

Light

Light makes it possible for us to see. We rely on vision more than any of our other senses.

Light travels as **waves** that move in a straight line. Some objects block or absorb light and cause a shadow to be created. Other objects, such as a mirror, reflect light. During **reflection**, light waves bounce off an object and change their direction of motion.

Light carries energy. When light is focused on an object with a magnifying glass, an intense amount of heat is generated. For instance, when light is focused on a small object, such as a leaf, it can begin to burn. Light can also be transformed into energy during photosynthesis, a process in which green plants use the energy of sunlight to manufacture food.

We grew up thinking of light as white. But if that is true, what are we seeing when we look at a rainbow? Even though the sunlight we see is white, it can be broken down into the many different colors of a rainbow.

You can also see the rainbow effect when light passes through a prism, a piece of glass whose ends are triangular. A prism also causes white light to be separated into a spectrum of different colors. This happens because light slows down and bends when it passes through the prism, an effect that occurs because glass is denser than air.

PRISM
(cross-section)

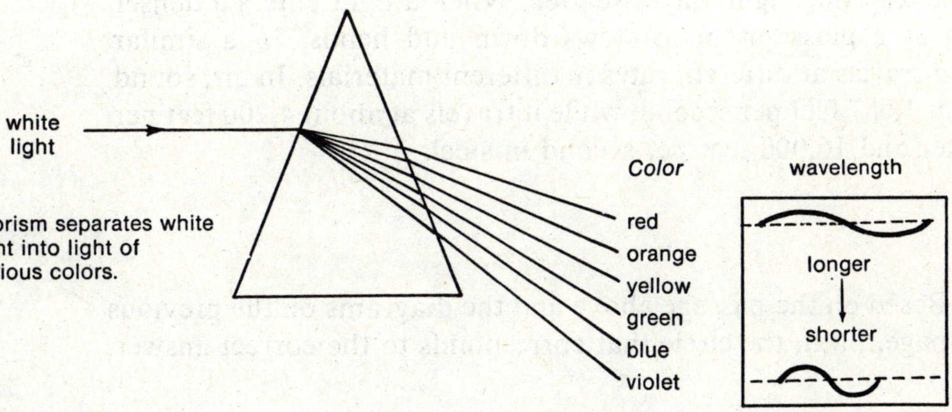

A prism separates white light into light of various colors.

The colors that make up white light are characterized by different wavelengths. Each wavelength of light is affected differently by its passage through a prism, the shortest wavelengths being the most affected. Violet light bends the most since it has the shortest wavelength. The colors with longer wavelengths bend the least. When we observe a rainbow, we are actually seeing white light that has been broken up, or **refracted**, into its many colors by its passing through raindrops.

THE FORMATION OF A RAINBOW

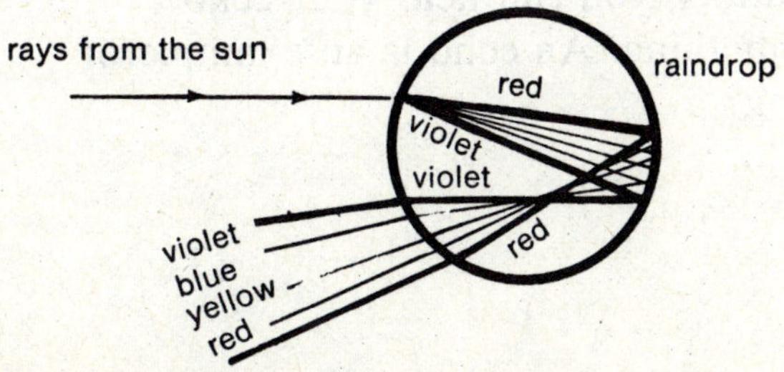

Sound

Sound has many of the characteristics of light. Sound is caused by vibrations. When an object vibrates it makes the air around it vibrate. These vibrations travel as **waves** through the air to our ears, and we hear them as sound. The term **pitch** refers to how rapidly the waves are generated. A high pitch will have a greater number of vibrations and will create a greater number of sound waves. The loudness of a sound is the strength of the sensation received by the ear.

The waves that create sounds and light have similar properties. Just as light is reflected by a mirror, sound can be reflected by a hard surface. A common example of the reflection of sound waves is when an echo is created by bouncing sound waves off a canyon wall. Sound waves scatter in much the same way that light waves scatter. When a light enters a denser material such as a glass prism, it slows down and bends. In a similar manner, sound travels at different rates in different materials. In air, sound travels at about 1,087 feet per second, while it travels at about 4,700 feet per second in water and 16,000 feet per second in steel.

Directions: Based on the passage above and the diagrams on the previous page, fill in the circle that corresponds to the correct answer.

18. A glass prism separates white light into its various colors because

(1) the different particles of light react with the glass
(2) the different wavelengths of light bend at different angles as they pass through the glass
(3) certain colors of light do not pass through glass and must bend around it
(4) glass is dense, so it consumes the energy of light
(5) glass is not dense, so the energy of light is used up

18 ① ② ③ ④ ⑤

19. When you are in a tunnel, you can hear your echo whenever you say something. An echo is an example of which of the following?

(1) Pitch
(2) Loudness
(3) Wavelength
(4) Refraction
(5) Reflection

19 ① ② ③ ④ ⑤

20. According to the diagram of the prism on page 173, which of the following has the longest wavelength?
(1) Red
(2) Orange
(3) Yellow
(4) Green
(5) Violet

20 ① ② ③ ④ ⑤

For answers and explanations, see page 182.

PHYSICS EXERCISE 3

Lasers

Laser light is very different from the incandescent light of a light bulb. Light bulbs emit many different wavelengths that move in many different directions. Laser beams are monochromatic, or of a single wavelength. They do not spread, even after traveling for miles.

At first, lasers were used only for scientific research. One experiment used laser beams to measure the distance from the earth to the moon to within half a foot. Using lasers, scientists have found that light travels at 186,282 miles per second. Some scientists are attempting to locate intelligent life on other planets by sending messages with lasers.

Lasers have been used in many practical ways. Lasers are used to weld and cut metals, resulting in better bonding of welds and cleaner cuts than can be achieved with conventional techniques. Lasers are used to cut fabric for clothing and drill holes in aerosol valves. Cables of glass fibers carrying laser beams may one day replace telephone cables.

Lasers have numerous applications in medicine. Surgeons use laser scalpels to perform delicate surgery and treat eye diseases. Lasers cauterize the blood vessels as they cut, a procedure that eliminates the bleeding that usually accompanies surgery. Some doctors claim that lasers will one day help people avoid heart bypass surgery by vaporizing fatty deposits in arteries.

The great advances of laser technology are now being put to uses that are less beneficial to humanity. On March 23, 1984, President Ronald Reagan created a stir in the scientific community by advocating the use of laser weapons. The possibility of "death ray" weapons has captured the public's imagination, but because of the secrecy that shrouds military projects, it is still a subject of speculation. The "smart bombs" that were used in Vietnam were guided by lasers, and many tanks are now equipped with laser range finders.

21. Which of the following is a characteristic of lasers? **21** ① ② ③ ④ ⑤
(1) Laser beams are used only for scientific research.
(2) Lasers emit beams with many different wavelengths.
(3) Laser beams travel long distances without spreading.
(4) Lasers have no practical applications.
(5) Laser beams scatter in many different directions.

22. Which of the following applications of laser technology would the author regard as not being beneficial to humanity? **22** ① ② ③ ④ ⑤
(1) Treatment of eye diseases
(2) "Smart bombs"
(3) Scientific research
(4) Transmittal of information
(5) Industrial applications

For answers and explanations, see page 182.

ELECTRICITY AND MAGNETISM

KEY WORDS

conductor—a material that easily transmits or carries electricity

Electrical Charges

Today many of us dry our clothes with antistatics like ClingFree or Bounce to keep clothes from sticking to each other. We know certain materials have a tendency to pick up electricity from the dryer and stick together. The same thing happens when a comb is rubbed against a cloth. The comb becomes electrified and causes hair to stand up.

Scientists have learned a great deal about electrical charges, which can either attract objects or repel them. Electrical charges can be either positive or negative. Objects that have the same charge repel each other, whereas objects with opposite charges are attracted to each other. Since every electrical charge is surrounded by an electrical field, the forces are stronger when the objects are moved toward each other. The repelling and attracting are the result of an interaction of electrical fields.

Magnets have similar characteristics. A magnetic object has the ability to attract uncharged iron objects. Magnets have two centers of magnetic force called the north and south poles. As with electricity, unlike magnetic poles attract and like poles repel each other. Magnets are surrounded by a magnetic field which decreases in strength as you move farther from the magnet.

Electric Current

One important characteristic of electricity is the way it moves or flows. This flow is known as *electric current*. Electric current has been compared to a stream of water. An electric wire offers resistance to the flow of electricity, similar to the way that long and narrow pipes resist the flow of water. Resistance to the flow of electric current depends upon the nature of the metal **conductor** and length of the wire. The force of the current is measured in volts. The current itself is measured in *ampres*, and resistance is measured in *ohms*.

Resistance to electric current in wires causes the temperature in a wire to rise, which further increases the resistance. If the amount of current passing through a wire is greater than the capacity of a wire, the heat from the resistance could start a fire. Electrical wiring systems in homes and buildings are protected by circuit breakers and fuses that automatically shut off the current if the wires become overloaded. To prevent accidental shocks, wires are covered with an insulator, a material such as plastic that will not conduct electricity.

For electricity to do useful work, it must be able to flow in a continuous path, or circuit. From the source of electrical energy, a battery or generator, the current flows through wires to an output device, such as a light or motor. To complete the circuit, a second wire must connect the output device to the source. If the circuit is broken, electricity will stop flowing through the wires and the output device will go off.

A SIMPLE CIRCUIT

Directions: Based on what you have just read, fill in the circle that corresponds to the correct answer.

23. Which of the following protects an electrical wiring system from an overload?
(1) Resistance
(2) Circuits
(3) Insulators
(4) Current
(5) Circuit breakers and fuses

23 ① ② ③ ④ ⑤

24. What is the unit of measure for resistance?
(1) Ohms
(2) Pressures
(3) Amperes
(4) Conductors
(5) Volts

24 ① ② ③ ④ ⑤

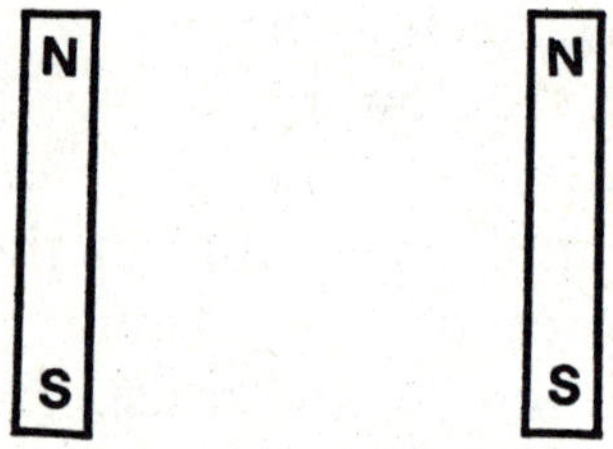

25. Which of the following is the most likely reaction when two bar magnets are placed side by side as shown in the diagram?
(1) The magnets would move away from each other.
(2) The magnets would move toward each other.
(3) Neither magnet would move.
(4) The poles on the left magnet would reverse themselves, but the magnets would not move.
(5) The poles on the right magnet would reverse themselves, but the magnets would not move.

25 ① ② ③ ④ ⑤

Use the following diagram to answer question 26.

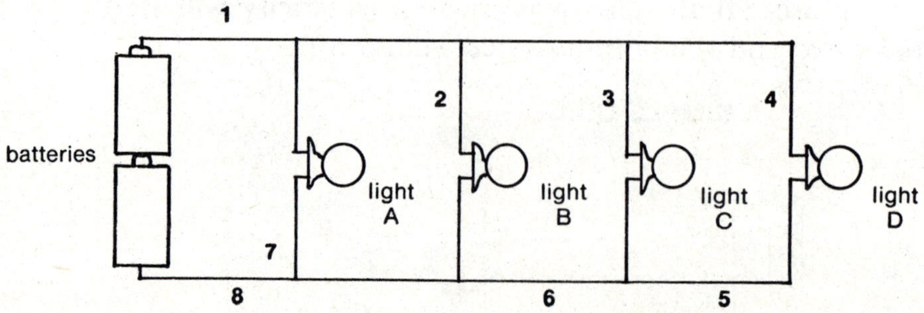

26. Which of the following wires should be removed to turn off only light D?

(1) Wire 1
(2) Wire 2
(3) Wire 3
(4) Wire 5
(5) Wire 8

26 ① ② ③ ④ ⑤

For answers and explanations, see page 182.

PHYSICS AND THE FUTURE

Humans can be distinguished from animals by the fact that they not only want to know what tomorrow will bring but also are driven by a desire to control the future. While animals are content with their day-to-day existence, seeking food, shelter, and safety, we plan and alter our world. Great curiosity and a desire to alter our world have led to many advances in physics and technology.

We try to predict the weather, however inaccurately, through daily weather reports. We attempt to design and develop tools that will improve our predictions. Many ideas in movies of the fifties and sixties that depicted the future have become reality. We have reached the moon. We have begun to conquer space. Computers and videotapes have allowed us to analyze the physics of sports to the point that men and women can run faster, jump higher, and throw farther.

We are in the midst of a scientific revolution. More than 90 percent of all scientists who have ever lived are alive and working today. Theories that were once widely accepted are being shaken to their roots. Scientists are trying to redefine matter and the laws that govern our solar system. Black holes appear to be a reality, the landing on Mars is history, and soon our satellites will have a direct view of what lies beyond our solar system. The popularity of films like *Star Wars*, *Star Trek*, and their numerous sequels attest to the fact that even our youth are fascinated by what may lie ahead.

Advances in electronic technology are rapidly changing our lives. Video games and home computers have transformed the recreation and business habits of millions of Americans. Who could have foreseen the miniaturization of electronic circuits? Today, thousands of circuits can be put on a tiny silicon chip, called a *microprocessor*, a device that is smaller than a thumbnail. For a relatively low price, a person can now buy a computer with more capacity than the best that were available only thirty years ago. Remember the $200 hand-held calculators of the early 1970s? Better and smaller ones can now be purchased for under $10. Great advances in physics and technology have revolutionized our lives. Who can imagine what the future will bring?

Directions: Based on what you have just read, fill in the circle that corresponds to the correct answer.

27. Futuristic movies of the fifties and sixties depicted

27 ① ② ③ ④ ⑤

(1) black holes
(2) fictional ideas that are now reality
(3) video games and home computers
(4) the physics of sports
(5) accurate weather reporting

28. According to the author, a major factor contributing to the electronic revolution is the

28 ① ② ③ ④ ⑤

(1) $200 calculator
(2) movie *Star Wars*
(3) television
(4) silicon chip microprocessor
(5) video game

29. The author mentions one difference between humans and animals. Which of the following answers would best describe the difference discussed in this passage?

29 ① ② ③ ④ ⑤

(1) Humans have begun to conquer space while animals seem to be content with planet earth.
(2) Humans are driven by a desire to control the future while animals seem content with their daily existence.
(3) Humans eat meat while all animals eat plants.
(4) Humans walk upright while animals all walk on four legs.
(5) Humans construct homes for shelter while animals do not build shelters.

For answers and explanations, see page 182.

ANSWERS AND EXPLANATIONS

INTRODUCTION

Finding Details	**1.**	(3)	The first line of this passage states that physics studies matter, energy, force, and time. The other choices are not discussed.
Making Inferences	**2.**	(1)	The passage tells you that through the study of optics, objects that use lenses have been invented. You can infer that a knowledge of optics would be useful in an investigation of material for eyeglass lenses.
Making Inferences	**3.**	(4)	The engineer would be concerned about the forces the jet body will encounter and should do research in the area of mechanics.

THE LAWS OF MOTION

Applying Ideas	**4.**	(4)	Newton's first law of motion is stated in the second paragraph. A moving car is a body in motion. Outside forces must be applied to stop it.
Making Inferences	**5.**	(1)	The third paragraph discusses acceleration. The greater the mass of the object to be moved, the slower the acceleration. The greater the force exerted upon the object, the greater the acceleration. The example shows that the same force exerted on a small and a large object will result in the smaller object moving more quickly. You can infer that *acceleration* means "increasing speed."
Making Inferences	**6.**	(3)	*Identify* means "recognize." The physical behavior of objects already existed. Sir Isaac Newton recognized the behavior, explained it, and demonstrated that these laws characterized objects in motion.
Drawing Conclusions	**7.**	(5)	The first paragraph tells you that there is no friction in a vacuum. You can conclude that a pendulum would behave differently inside a vacuum from the way it would behave outside the vacuum because there would be no friction from air molecules.
Applying Ideas	**8.**	(3)	Newton demonstrated that for every action there is an equal and opposite reaction. The action, the force of the escaping water, causes a reaction, the rotation of the sprinkler in the opposite direction.

PHYSICS EXERCISE 1

Finding Details	**9.**	(5)	Lines 11–14 state that, by using the wind tunnel simulations, the downhill racers intend to reduce drag.
Main Idea	**10.**	(5)	In order to improve their performance in the Olympics, the skiers are using the wind tunnel to simulate racing conditions.
Applying Ideas	**11.**	(2)	The diagram shows by arrows pointing in the opposite direction that the force that works against the forward motion of the ski jumper is drag.

STATES OF MATTER

Finding Details	**12.**	(5)	The first paragraph describes the three states of matter.

Applying Ideas **13.** (1) Granules of instant coffee are an example of a solid. They will not change shape, and they have a constant volume. Of course, if you add water to the granules, the state of matter quickly changes.

Finding Details **14.** (2) The fourth paragraph states that, if a gas is sealed in a container, the gas takes on the shape of that container.

PHYSICS EXERCISE 2

Finding Details **15.** (1) The third paragraph mentions that freon is usually used.

Drawing Conclusions **16.** (3) The second paragraph states that, when a liquid evaporates, heat is absorbed by the gas.

Finding Details **17.** (5) The third paragraph says that, in the condenser, heat is released into the air.

LIGHT AND SOUND

Finding Details **18.** (2) The essay states that the different types of light that make up white light have different wavelengths and bend differently. This creates the spectrum of different colors.

Applying Ideas **19.** (5) The essay mentions in the last paragraph that echoes are created by sound reflecting off a hard surface.

Applying Ideas **20.** (1) The diagram shows the colors of the spectrum and their relative wavelengths. Red is the color with the longest wavelength.

PHYSICS EXERCISE 3

Finding Details **21.** (3) Lines 4–5 mention that laser beams can travel for miles without spreading. None of the other choices are found in the passage.

Making Inferences **22.** (2) The general tone of the last paragraph indicates that the author does not regard "smart bombs" as beneficial to humankind.

ELECTRICITY AND MAGNETISM

Finding Details **23.** (5) The next to the last paragraph mentions that the electrical wiring systems in homes and buildings are protected by circuit breakers and fuses.

Finding Details **24.** (1) The fourth paragraph states that resistance is measured in ohms.

Applying Ideas **25.** (1) The third paragraph states that like poles repel each other. The most reasonable response would be that the magnets would move away from each other.

Applying Ideas **26.** (4) The last paragraph mentions that, if a circuit is broken, electricity will stop flowing through the wires and the device will go out. Wires 1 and 8 lead to all the bulbs, so, if they are removed, all the bulbs will go out. Lights A, B, and C are fed by wires 7, 2, and 3 respectively. The wires that go to Light D only are 4 and 5.

PHYSICS AND THE FUTURE

Finding Details **27.** (2) The passage tells you that many ideas in movies of the fifties and sixties that depicted the future have become a reality.

Making Inferences **28.** (4) The tiny silicon chip microprocessor has made most of the electronic advances possible.

Finding Details **29.** (2) The first paragraph states the difference between humans and animals.

SCIENCE POST-TEST

Introduction

This post-test is similar to the GED science test. On both tests, you will read science materials and answer questions combining your knowledge of science with the reading skills you have learned. If you have worked through the exercises in this book, you should be ready to do the post-test.

You should take approximately 90 minutes to complete this test. At the end of 90 minutes, stop and mark your place. Then, go on and finish the test. This will give you an idea of whether or not you can finish the real GED test in the time allotted. Try to answer as many questions as you can. A blank will count as a wrong answer, so make a reasonable guess for questions you are not sure of.

You will need to go through the discrete items first, the items that test your science knowledge. Do the best that you can on these and then move on to the other areas. This is the hardest section of the test for most people because it tests their recollection of certain scientific facts and concepts.

Work as quickly and as carefully as you can. Try to answer every question. Answers that are left blank are automatically counted wrong, on both this test and the actual GED test. If a question seems too difficult or you are unsure of an answer, make an educated guess instead of leaving it blank.

Use the answer grid on page 201 to record your answers. Be sure to fill in only one circle for each question, or it will be marked incorrect. When the time limit of ninety minutes has passed, make a mark on your answer grid to indicate how far you got on the post-test and then complete the test. This will give you an indication of how much of the test you can complete in the allotted time and will also measure how well you do on the science material.

Use the post-test evaluation chart on page 200 to determine your readiness for the actual GED test. The scoring chart will tell you the relationship between the number of correct answers and certain GED scores.

Part A

Directions: Each of the questions or incomplete statements below is followed by five possible answers. Choose the one that is the best and fill in the corresponding circle on the answer grid.

1. Which of the following is the most common blood type among North Americans?
(1) O
(2) A
(3) B
(4) AB
(5) AO

2. DNA can be best described as
(1) denatured alcohol
(2) a nucleic acid that forms chromosomes
(3) the living tissue of a plant
(4) the Doctor and Nurse Association
(5) the material made by the genes

Question 3 refers to the following diagram.

PEA PLANTS

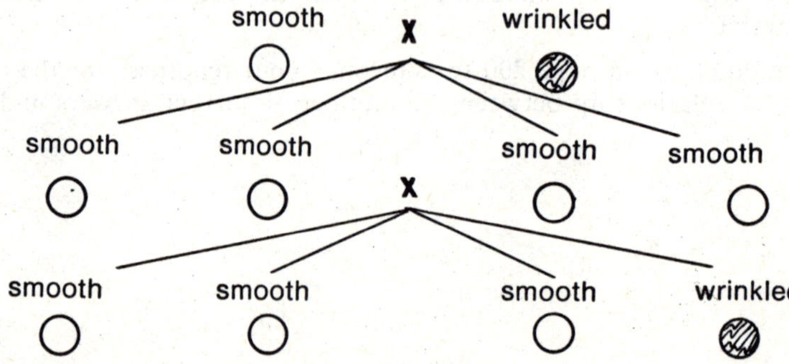

3. The above diagram shows the result when a smooth pea plant is crossed with a wrinkled pea plant. From the information in the diagram, you could conclude that the smooth trait is
(1) mutated
(2) recessive
(3) dominant
(4) hidden
(5) incomplete

GO ON TO NEXT PAGE.

4. The pollination of a plant is a function of
(1) digestion
(2) photosynthesis
(3) symbiosis
(4) reproduction
(5) oxidation

5. Heavy cigarette smoking, obesity, and stress are major factors causing which of the following?
(1) Leukemia
(2) Hypertension
(3) Anemia
(4) Emphysema
(5) Bronchitis

6. A function of the umbilical cord is to
(1) carry food from the placenta to the embryo
(2) store certain fats for future use
(3) carry wastes from the placenta to the embryo
(4) hold the fetus to the uterine wall
(5) mix the maternal and fetal blood

7. Testicles are to sperm as ovaries are to
(1) babies
(2) birth control
(3) eggs
(4) the uterus
(5) reproduction

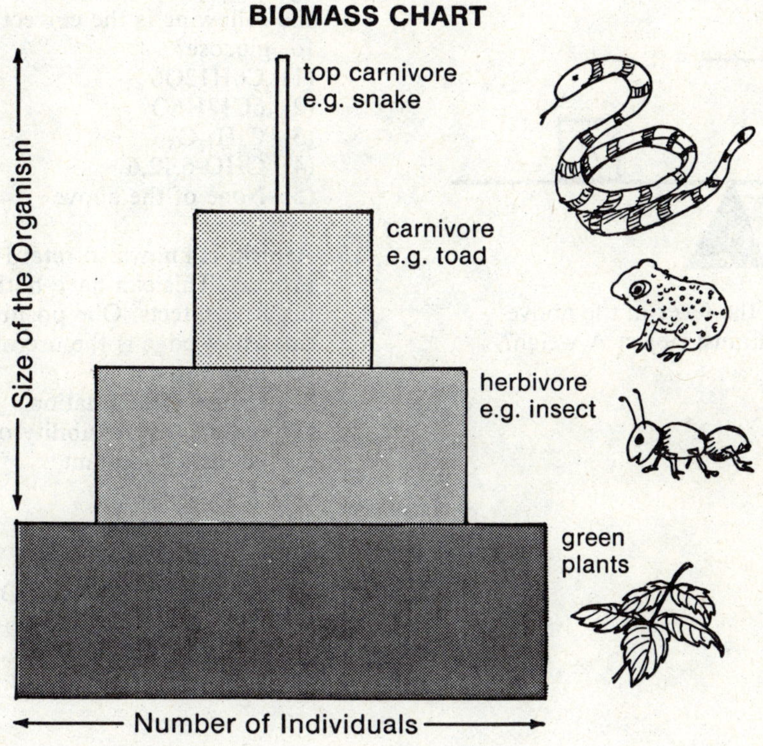

8. According to the biomass chart shown above, which segment of the biomass is the most populous?
(1) Top carnivores
(2) Carnivores
(3) Herbivores
(4) Green plants
(5) None of the above

GO ON TO NEXT PAGE.

9. You are on a beach in Hawaii and hear on the radio that an extremely severe earthquake has just struck, centered off the southern coast of Alaska. Which of the following would be the most prudent course of action?
(1) Look out for a volcanic eruption in the islands.
(2) Be alert for crevices in the sand.
(3) Prepare your surfboard for the greatest waves you've ever seen.
(4) Leave the beach area because of a possible tsunami.
(5) Be prepared for a sudden drop in temperature.

10. The earth's atmosphere is composed mainly of
(1) hydrogen
(2) oxygen
(3) argon
(4) nitrogen
(5) helium

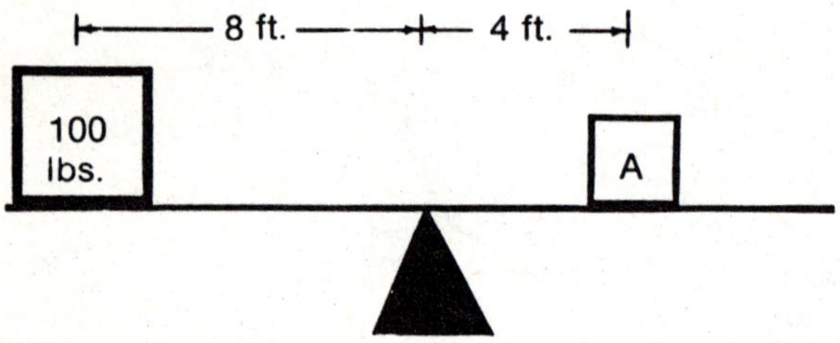

11. In order to balance the lever in the above diagram, how much must object A weigh?
(1) 50 pounds
(2) 100 pounds
(3) 200 pounds
(4) 500 pounds
(5) 1,000 pounds

12. If a gas has a volume of one liter, and the gas is compressed until it occupies only 750 cubic centimeters, which of the following is most likely?
(1) The pressure will increase.
(2) The pressure will decrease.
(3) The temperature will decrease.
(4) The volume will increase.
(5) The molecules of oxygen will stop moving.

13. Which of the following represents the three states of matter?
(1) Water, vinegar, and vegetable oil
(2) Rocks, water, and sand
(3) Salt, pepper, and sugar
(4) Air, helium, and carbon dioxide
(5) Ice, water, and water vapor

14. Photosynthesis produces glucose, which contains six atoms of carbon, twelve atoms of hydrogen, and six atoms of oxygen. Which of the following is the correct empirical formula for glucose?
(1) C6H12O6
(2) 6C12H6O
(3) $C_6H_{12}O_6$
(4) CHO-6,12,6
(5) None of the above

15. Aspirin is known to retard the blood-clotting process. This can have both positive and negative effects. One positive application of this knowledge is the use of aspirin to
(1) relieve pain
(2) reduce inflammation
(3) reduce the possibility of a stroke
(4) act as a stimulant
(5) stop bleeding

GO ON TO NEXT PAGE.

16. As the two bar magnets below are brought closer together, what will happen?

(1) They will attract each other.
(2) They will repel each other.
(3) They will cancel each other's magnetic field.
(4) They will create alternating current.
(5) Nothing will happen.

17. The gradual destruction of the ozone layer is regarded as undesirable since the ozone layer protects us from
(1) acid rain
(2) harmful rays from the sun
(3) infections
(4) pollution of factories
(5) meteorites

18. If you have 1,000 grams of a radioactive material that has a half-life of 10,000 years, how much of that material will be left after 20,000 years?
(1) 1,000 grams
(2) 500 grams
(3) 375 grams
(4) 250 grams
(5) 125 grams

19. A fossil is
(1) a type of igneous rock that is usually found below the earth's surface
(2) a plant hormone that helps promote growth
(3) the study of the relationship of living things and their environment
(4) a fracture in the earth's crust
(5) the remains of an animal or plant preserved in a rock

GO ON TO NEXT PAGE.

Part B

<u>Directions:</u> Each passage in this section is followed by questions. After reading a passage, choose the best answer to each question and fill in the corresponding circle on the answer grid.

<u>Questions 20–24</u> refer to the following passage.

Visible light is the light that people are able to see, such as sunlight or candlelight. It is a small part of the electromagnetic spectrum that includes X rays, ultraviolet rays, visible light, and radio waves.

While humans cannot see the invisible part of the spectrum, there are numerous ways to use it. X rays are used by doctors to treat cancer and to locate disorders inside the body. Ultraviolet rays are used in sunlamps. Infrared rays are used to treat skin diseases and to dry paint on cars. Radio waves, like the one shown below, are used in radio and television broadcasting.

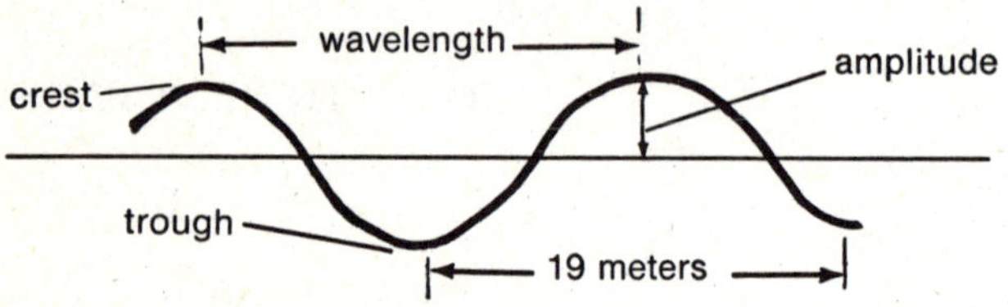

The differences among the colors of the spectrum lie in their wavelengths. A wavelength is measured by the distance between the crests or troughs of a wave. Wavelengths range from less than 10 trillionths of a meter (gamma rays) to more than 10,000 kilometers (long radio waves). Visible light has a wavelength between 400 and 730 millimicrons (there are 1,000,000 millimicrons in a millimeter).

The visible light spectrum, or white light, can be broken down into its component colors, as in a rainbow. Although light coming from the sun looks colorless, it is a mixture of different wavelengths ranging from the shortest wavelengths of violet to the longest wavelengths of deep red.

20. The shortest wavelength visible to man is
(1) red
(2) violet
(3) orange
(4) yellow
(5) blue

21. You can conclude from the passage that the length of electromagnetic waves
(1) is always long
(2) is unimportant to human vision
(3) is determined by pigment
(4) is always the same
(5) varies greatly

22. The radio wave shown in the diagram is used widely in international shortwave radio broadcasting. The label "19 meters" represents the
(1) wavelength
(2) crest
(3) trough
(4) frequency
(5) amplitude

23. Which of the following types of rays is <u>not</u> mentioned as being part of the electromagnetic spectrum?
(1) Radio waves
(2) Cosmic waves
(3) Ultraviolet waves
(4) Infrared waves
(5) X rays

24. When light passes through a glass prism, it emerges and displays the colors of a rainbow. This is due to which of the following?
(1) The ultraviolet rays disrupt the visible light waves.
(2) The pigment of the glass absorbs the lighter colors.
(3) White light is broken down into its different component colors.
(4) White light is absorbed by the prism, leaving only the colors.
(5) The light creates an optical illusion.

GO ON TO NEXT PAGE.

Questions 25–27 refer to the following chart.

TRANSFUSION RELATIONSHIP

Blood Type	*Can Receive Blood from*	*Acts as Donor to*
O	O	O, A, B, AB
A	O, A	A, AB
B	O, B	B, AB
AB	O, A, B, AB	AB

25. What type(s) of blood can be given to a person with blood type A?
(1) Only type O
(2) Only types O and A
(3) Only types O and B
(4) Only types A and AB
(5) Any blood type

26. If a person with blood type AB wishes to donate blood, those with which blood type(s) may receive the blood?
(1) Only type O
(2) Only types O and B
(3) Only types B and AB
(4) Only type AB
(5) All types of blood

27. Which of the following blood types can be regarded as a universal donor?
(1) O
(2) A
(3) B
(4) AB
(5) All blood types

GO ON TO NEXT PAGE.

Questions 28–32 refer to the following passage.

The effects of smoking on the unborn fetus are not good. The fetus of a mother who smokes two packs a day loses 40% of the oxygen that is normally present. Lack of the oxygen needed to grow at a normal rate may explain why babies born to smokers average six to seven ounces lighter at birth than the babies of nonsmokers.

Lower birth weight is not the only adverse effect of smoking on babies and their mothers. Women who smoke less than a pack a day increase the chances of their babies dying by 20%; those who smoke more than a pack a day are 36% more likely to have their babies die. Smokers are twice as likely to have miscarriages and will also suffer more complications during the pregnancy, more premature births, and a greater chance of having a stillbirth.

Even though a woman stops smoking before delivery, smoking during pregnancy makes it more likely her baby will have malformations of the heart or other organs. Smoking during pregnancy doubles the chances of abruptio placentae—a premature separation of the placenta from the uterine wall that may result in the death of the fetus.

Difficulties do not end at birth. The children of smokers are 52% more likely to die of crib death and twice as likely to develop a lung illness, such as bronchitis or pneumonia.

28. On the average, how much lighter are newborn babies of women who smoke during pregnancy than the newborns of nonsmokers?
(1) No difference
(2) One to two ounces
(3) Two to three ounces
(4) Three to four ounces
(5) Six to seven ounces

29. What is abruptio placentae?
(1) A minor complication, occurring only during the first trimester of pregnancy
(2) A birth defect involving an underdeveloped digestive tract
(3) A condition in which the fetus receives a diminished supply of oxygen
(4) Another term for Sudden Infant Death Syndrome, or crib death
(5) Premature separation of the placenta from the uterine wall

30. Which of the following does the author cite as the most likely reason for a lower birth weight among infants of mothers who smoke?
(1) An insufficient supply of oxygen to the fetus
(2) The presence of nicotine in the mother's blood
(3) The mother's poor eating habits
(4) Sudden Infant Death Syndrome
(5) Mothers who smoke suffer more complications

31. Which of the following is not mentioned as a possible adverse effect of a pregnant woman's smoking?
(1) Increased risk of miscarriage
(2) More premature births
(3) Severe brain damage
(4) Greater possibility of a stillbirth
(5) Lower birth weight

32. How much greater are the risks of infant death for pregnant women who smoke more than a pack a day than the risks for pregnant women who smoke less than one pack a day?
(1) The risks are the same
(2) 16 percent
(3) 20 percent
(4) 36 percent
(5) 40 percent

GO ON TO NEXT PAGE.

Questions 33–36 refer to the following passage.

Man has long marveled at lightning. It was the ultimate weapon of the gods of ancient civilization. Today, lightning is subjected to scientific scrutiny, but it is no less awesome, and it still deserves respect.

National Center for Health Statistics data for recent years show that lightning kills about 125 Americans per year and injures more than 500. Property loss is estimated in the hundreds of millions of dollars annually.

Lightning is an effect of electrification within a thunderstorm. As the thunderstorm develops, interactions of charged particles produce an intense electrical field within the cloud. A large positive charge is usually concentrated in the frozen upper layers of the cloud, and a large negative charge along with a smaller positive area is found in the lower portions.

The earth is normally negatively charged with respect to the atmosphere. But as the thunderstorm passes over the ground, the negative charge in the base of the cloud induces a positive charge on the ground below and for several miles—km—around the storm. The ground charge follows the storm like an electrical shadow, growing stronger as the negative cloud charge increases. The attraction between positive and negative charges makes the positive ground current flow up buildings, trees, and other elevated objects in an effort to establish a flow of current. But air, which is a poor conductor of electricity, insulates the cloud and ground charges, preventing a flow of current until huge electrical charges are built up.

Lightning occurs when the difference between the positive and negative charges—the electrical potential—becomes great enough to overcome the resistance of the insulating air and to force a conductive path for current to flow between the two charges. Potential in these cases can be as much as 100 million volts. Lightning strokes proceed from cloud to cloud, cloud to ground, or, where high structures are involved, from ground to cloud.

33. What inhibits the flow of electricity between the massed negative charge in the base of the storm cloud and the positive charges on the ground?
(1) Air
(2) Water vapor
(3) Raindrops
(4) Step leaders
(5) Conductive paths

34. What causes the positive charges on the ground to build up?
(1) The raindrops are charged as they fall to the ground.
(2) Electrical ions are carried downward by the cold winds that accompany thunderstorms.
(3) Pilot leaders, which are too faint to be visible, carry electrical ions.
(4) Conductive channels between the clouds and the ground carry electrical charges.
(5) There is an attraction between the negative ions in the cloud and the positive ions on the ground.

35. About how long does a typical cloud-to-ground stroke last?
(1) Instantaneous
(2) One millisecond
(3) One second
(4) Four seconds
(5) Ten seconds

36. If you were caught in a thunderstorm, which of the following would be the safest?
(1) Lying in a drainage ditch
(2) Standing under a tall tree
(3) Sitting in a boat on a large lake
(4) Walking along a railroad track
(5) Standing in an open field

The typical cloud-to-ground stroke begins as a pilot leader, too faint to be visible, advances downward from the cloud, setting up the initial portion of the stroke path.

A surge of current called a *step leader* follows the pilot, descending 100 feet—30 meters—or more at a time, pausing, then moving on until the conductive path of electrical (ionized) particles is near the ground.

There, discharge streamers from the ground, intercept the leader path and complete the conductive channel between ground and cloud charges.

When this path is complete, a return stroke leaps upward, illuminating the branch of the descending leader track. Because these tracks point downward, the stroke appears to come from the cloud. A typical cloud-to-ground stroke takes about one second.

GO ON TO NEXT PAGE.

Questions 37–41 refer to the following passage.

Hope is running out for the survival of the dusky seaside sparrow species. The four birds in captivity at Walt Disney World in Lake Buena Vista, Florida, are all male. No female has been seen since 1975, and no males have been seen in the wild since 1980. In an attempt to save at least some of the dusky sparrow's characteristics, Disney biologists plan to attempt to breed the four males with seaside sparrows of other species. These experiments could eventually produce a hybrid sparrow with most of the dusky's genes. The black-and-white canary-size dusky, weighing less than an ounce, spent most of its time darting about the ground in its limited floodplain and tidal marsh range. The dusky's habitat was destroyed first by mosquito abatement measures, then by human population growth associated with the space program and a highway built to facilitate access to Walt Disney World, and finally by range fires. The dusky seaside sparrow is expected to be the first vertebrate known to become extinct since the federal Endangered Species Act was passed in 1973.

37. Which of the following is not cited as a contribution to the extinction of the dusky sparrow?
(1) Human population growth
(2) Hybrid sparrows
(3) A highway to Walt Disney World
(4) Range fires
(5) Mosquito abatement measures

38. What measures are biologists taking to try to save some of the dusky sparrow's characteristics?
(1) Reconstruct the floodplain
(2) Search for female dusky seaside sparrows
(3) Artificially inseminate female dusky sparrows
(4) Tear down the highway to Walt Disney World
(5) Breed duskies with sparrows of other species

39. Which of the following would be a good title for this passage?
(1) The Endangered Species Act
(2) New Zoo Procedures
(3) The Last Duskies in Florida
(4) The Dusky Sparrow's Habitat
(5) Walt Disney World

40. What is the most likely source of food for the dusky sparrow?
(1) Mosquitoes
(2) Worms
(3) Beetles
(4) Minnows
(5) Mice

41. The dusky sparrow's habitat has undergone multiple transformations. Using information from the passage and what you already know about ecosystems, infer how these transformations have affected other organisms in that area.
(1) It is unlikely that any of the other organisms were affected by the changes.
(2) It is unlikely that any of the other organisms were more than minimally affected.
(3) The changes have affected all of the other organisms in varying degrees but not to the extent that they have affected the dusky.
(4) The other types of seaside sparrow will probably suffer the same fate as the duskies.
(5) The dusky is just the first to become extinct; others will follow.

GO ON TO NEXT PAGE.

Questions 42–46 refer to the following passage.

The advent of synthetic shampoos in the past several decades has, of course, revolutionized cleansing and cleaning chores, whether household, industrial, automotive, or cosmetic. Synthetic detergents are classified as anionic, cationic, nonionic, or amphoteric, depending on the way the ions behave in water. To some degree, all have the capacity to break down the resistance barrier between the water and the dirt, oil, or other material on the surface to be cleaned and allow the material to be rinsed away. Synthetic detergent shampoos sold for adults are usually of the anionic type; for babies and children, of the amphoteric type or amphoteric mixed with the anionic type. Cationic synthetic detergents are not usually included in shampoos as detergent agents.

The major advantage of synthetic detergents over soap in shampoos is their effective functioning in hard water. Soap, an alkali salt of fatty acids, works well in removing dirt and grease from surfaces, but in hard water, which contains a relatively high amount of calcium in solution, the calcium reacts with the soap to form deposits of a gummy material called soap scum. The familiar ring in the bathtub is composed of these deposits, to which dirt and other undissolved matter in the water may adhere. Such scum can form on the hair during shampooing with soap and hard water, dulling its lustre and making it difficult to comb unless rinsed with some substance which will redissolve the scum. Soap's foaming and cleansing action is reduced according to the degree of hardness of the water. Synthetic detergents do not react with the calcium in the water, and scum deposits do not form, even in the hardest water, though the detergent or cleaning action may itself vary depending on the kind of synthetic detergent used and the kind of foreign material to be removed. . . .

When soap was still the thing for hairwashing, the scum that formed to dull the hair's lustre was often reduced or eliminated by following up with a rinse of such acidic substances as vinegar and lemon juice.

42. Which of the following can be used to dissolve soap scum?
(1) Soap
(2) Fatty acids
(3) Lemon juice
(4) Alkali salts
(5) Undissolved materials

43. A manufacturer claims that the anionic synthetic shampoo that she produces is suitable for babies and small children. From the information in the passage, which of the following is probably the mildest type of synthetic cleansing agent?
(1) Anionic
(2) Cationic
(3) Nonionic
(4) Amphoteric
(5) No differences

44. According to the passage, which of the following is needed to dissolve the scum formed by synthetic detergents?
(1) Vinegar
(2) Protein derivatives
(3) Alkali salt
(4) Conditioner
(5) None of the above

45. In hard water, what is it that reacts with soap to form soap scum?
(1) Undissolved matter in the water
(2) Bathtub rings
(3) Dirt
(4) Vinegar
(5) Calcium

46. If you came across a bathtub with a film of an oily, yet gritty, substance containing particles of hair, you could conclude that the person who had just finished taking a bath had used which of the following to wash his/her hair?
(1) Soap
(2) An anionic type of synthetic shampoo
(3) A cationic type of synthetic shampoo
(4) A nonionic type of synthetic shampoo
(5) An amphoteric type of synthetic shampoo

GO ON TO NEXT PAGE.

Questions 47–53 refer to the following passage.

Radiation can kill or damage cells. If enough cells die, the organ they form will die; if crucial organs die, the organism will die. Thus, one consequence of radiation is death.

To be immediately lethal, radiation exposure to the whole body must exceed 1,000 rems over a brief period of time—minutes or hours (such exposures occurred at the Hiroshima and Nagasaki bombings). A dose of 500 rems, delivered at one time to the whole body, will cause death, on the average, in 50 percent of the cases.

In the range from 500 rems down to 100 rems, radiation sickness occurs and some individuals will die. At lower radiation levels, the consequences are more difficult to predict and detect.

For low radiation doses it is cell damage, not cell death, that is harmful. A dead cell can be replaced. Damage, however, can replicate itself and multiply. The type of damage depends on the nature of the struck cell.

If it is an ordinary cell—bone tissue or flesh, for instance—the damage is confined to the struck organism. This is called somatic damage. The most feared type of somatic damage is cancer or leukemia. If, however, a reproductive cell is damaged, then a mutation can be caused and the damage transmitted to future generations.

That radiation can cause cancer or genetic mutation is not questioned. What is questioned is the relationship between the dose—particularly doses below one rem—and the number of resulting cancers or mutations. For both types of damage, the latency period—the time between exposure and the effect—is long. It is 25 or so years for cancer and a generation or more for genetic damage.

47. In a very short period of time, what level of radiation exposure is considered to be lethal in 50 percent of the cases?
(1) 1,000 rems or more
(2) 500 rems
(3) 100 rems
(4) 50 rems
(5) below 1 rem

48. According to the passage, which of the following is the most feared type of somatic damage?
(1) Stomach disorders
(2) Cancer
(3) Radiation sickness
(4) Dead cells
(5) First degree burns

49. The most damaging consequence of low levels of radiation is
(1) under 100 rems
(2) death of an organ
(3) radiation sickness
(4) cell damage
(5) cell death

50. What is the latency period for genetic damage?
(1) A half-life
(2) A generation or more
(3) 7 years
(4) 150 years
(5) None of the above

51. A good title for this passage would be
(1) The Biological Effects of Radiation
(2) The Legacy of Hiroshima
(3) Somatic Diseases
(4) The Atomic Bomb
(5) The Nuclear Winter

52. What is the danger when a reproductive cell is damaged?
(1) The cell could die.
(2) Subsequent exposure to high levels of radiation could be lethal.
(3) There would be an increase in the incidence of multiple births.
(4) Genetic damage can be transmitted to future generations.
(5) The danger is no different from when other types of cells are damaged.

53. Why is it difficult to predict the long-term effects of low levels of radiation?
(1) There are no long-term effects of low levels of radiation.
(2) There are not enough people who have been exposed to low levels of radiation.
(3) The Pentagon has suppressed all of its data on radiation exposure.
(4) We have never had a thermonuclear war.
(5) The latency period can be 25 years for cancer and generations for mutations.

GO ON TO NEXT PAGE.

<u>Questions 54–55</u> refer to the following graph.

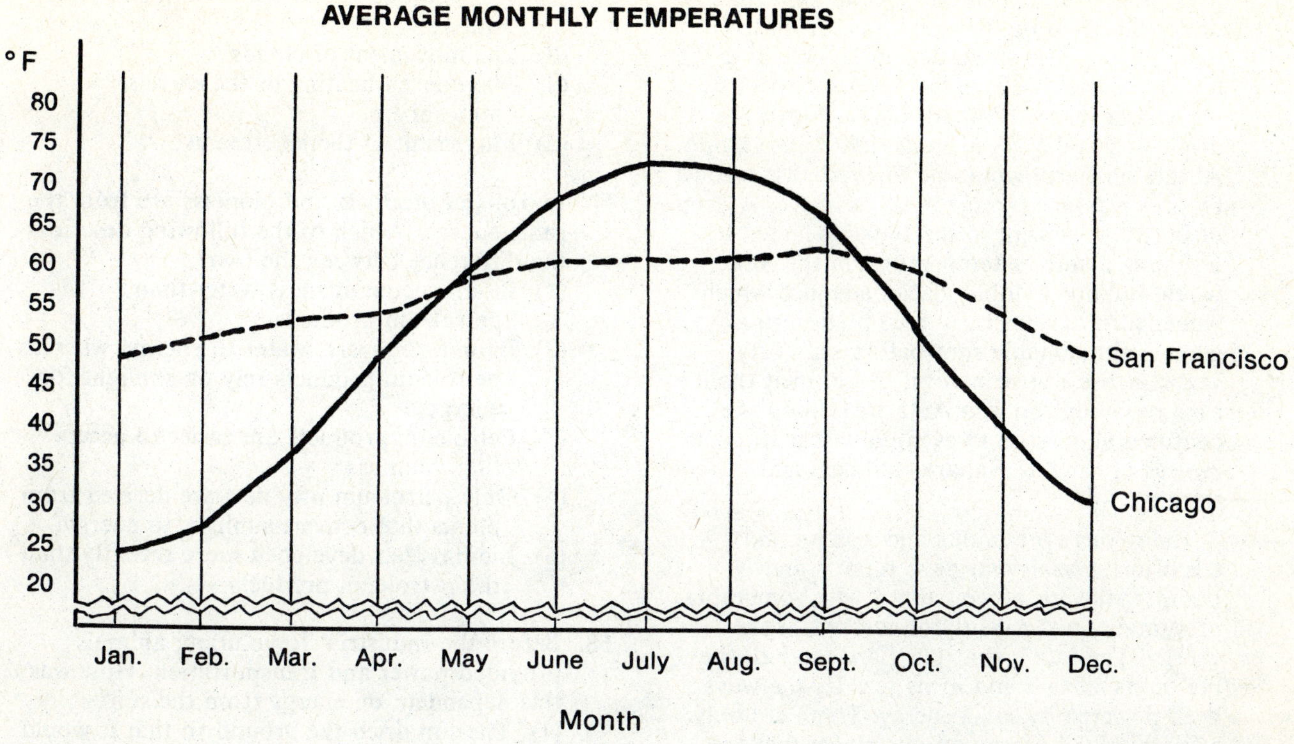

54. During which month do Chicago and San Francisco have almost the same average temperatures?
(1) March
(2) May
(3) July
(4) October
(5) November

55. Which of the following conclusions is based on information from this graph?
(1) Summers are generally hotter in San Francisco than Chicago.
(2) Chicago is warmer than San Francisco during the months of March, April, and October.
(3) The temperature extremes are greater in Chicago than in San Francisco.
(4) San Francisco has more severe winters than Chicago.
(5) Mild winters are more characteristic of Chicago than San Francisco.

GO ON TO NEXT PAGE.

Questions 56–60 refer to the following passage.

On Earth, all life of every kind from the tiniest viroid to the largest whale, from the blue-green algae to giant sequoia, including the human species itself, is fully and precariously dependent on the Sun. All life on Earth originated and evolved in its light and warmth. All life on Earth would be forever extinguished by only a minor change in its radiance. A drop of only two percent in the 10,000 degree Fahrenheit surface temperature of the Sun would initiate a global glacial advance which would turn our beautiful blue plant into a giant, uninhabitable snowball in only fifty years. A rise of two percent would melt the ice caps of Greenland and Antartica, flood the continental coasts and eventually sear the land masses of earth to Saharan stillness and sterility.

Today there are books and articles and television specials and news reports and bumper stickers, all concerned with something presumably just recently discovered—solar energy. The fact is that our Earth and all the life on its surface and in its seas have always been powered by solar energy. There are only a few kinds of energy on our planet that are not solar: e.g., geothermal (heat from the deep interior of the Earth); tidal (the ebb and flow of the tides caused by the gravitational pull of the Moon); and nuclear. Oil, gas, and coal all come from the fossils of plants which grew by converting sunlight to energy. Wood and "biomass" are from the same source but more recent. Hydroelectric power is possible only because the Sun evaporates the Earth's standing water and pours it back in rainfall into the sources of the rivers which drive the water turbines. Even the wind that turns our windmills—and modern windmills or "wind turbine generators" hold great promise as a clean, renewable energy source—are born out of the uneven heating of the Earth's atmosphere by the Sun. . . .

In ultimate confirmation of the closeness of their relationship, Earth and Sun will one day die together as they were born together. Some five billion years in the future the Sun will begin to exhaust the hydrogen fuel in its deep interior, and the fusion process will move outward, expanding the Sun into a red, giant star which will engulf and incinerate the inner planets, probably including the Earth. But there is no comfort in that "probably," because even if it is not physically swallowed by the swollen Sun, our planet will by then long have been a charred and lifeless cinder swinging residually through space.

56. What creates wind?
(1) The gravitational pull of the moon
(2) The evaporation of water from standing water
(3) The movement of clouds
(4) The uneven heating of the earth's atmosphere
(5) Movement of the jet streams

57. Petroleum products and biomass are from the same source. Which of the following describes the difference between the two?
(1) Biomass contains less water than petroleum products.
(2) Biomass converts water to energy, whereas petroleum products rely on sunlight for energy.
(3) Petroleum products are easier to access than biomass.
(4) Only petroleum products are derived from plants that convert sunlight to energy.
(5) Biomass has developed more recently than the petroleum products.

58. Before the Industrial Revolution, animals provided power and transportation. How was this dependent on energy from the sun?
(1) The sun dried the ground so that it would be dry enough to work.
(2) The uneven heating by the sun created breezes that cooled the laboring animals.
(3) Wood, which grew through the conversion of sunlight, was burned to keep the animals warm in the winter.
(4) The animals derived their energy from plants, which had used the sun's energy for their own growth.
(5) This was not dependent on the sun's energy.

59. If the surface temperature of the sun were to decrease to about 9,800° F, or 2 percent below its present temperature, which one of the following would be the most likely long-term effect?
(1) The level of the oceans would rise.
(2) The surface of the earth would be covered by ice masses.
(3) The surface of the earth would become desertlike.
(4) Desert regions would be transformed into lush, tropical rain forests.
(5) The average daily temperature would drop a degree or two.

60. What is geothermal energy?
(1) The heat from deep within the earth
(2) The fusion of minerals
(3) The conversion of plant matter to coal
(4) The movement of underground streams
(5) The ebb and flow of tides

END OF EXAMINATION

ANSWERS AND EXPLANATIONS

1. (1) The most common blood group among North Americans is type O.

2. (2) DNA (deoxyribonucleic acid) is a nucleic acid that is found in the nuclei of cells. DNA and protein form chromosomes, the bearers of genetic information. This transmits hereditary instructions from a parent cell to its offspring and makes it possible for an organism to produce another organism much like itself.

3. (3) Since the number of offspring having smooth skins outnumbers the number of offspring with wrinkled skins, you can conclude that the trait for smooth skin is dominant.

4. (4) The growth and spreading of pollen are a part of the reproductive process in plants.

5. (2) All three of these factors are related to hypertension, or high blood pressure.

6. (1) The umbilical cord carries nutrients from the placenta to the embryo. The fetus's blood does not mix with the mother's blood.

7. (3) Sperm are the male reproductive cells that mature in the testicles. Eggs are the female reproductive cells that mature in the ovaries.

8. (4) The label at the bottom of the biomass chart indicates that the width of each bar shows the number of individuals in that category. The longest bar is in the category labeled "Green Plants."

9. (4) Tsunami are huge sea waves caused by an earthquake under the ocean. Given the fact that there has just been a severe earthquake under the ocean, it would be advisable to leave the beach area and move inland.

10. (4) Approximately 78 percent of the earth's atmosphere is composed of nitrogen.

11. (3) If object A is half the distance from the middle of the lever as the 100-pound weight, then in order to balance, it must weigh twice as much, or 200 pounds.

12. (1) The volume of a gas varies inversely with the pressure on the gas. In other words, when the volume decreases, the pressure increases.

13. (5) The three states of matter are solid, liquid, and gas. Ice is a solid, water is a liquid, and water vapor is a gas.

14. (3) The correct empirical formula for glucose is $C_6H_{12}O_6$. Chemists indicate the number of atoms by writing the number after the letter, and a little below it.

15. (3) Since aspirin retards clotting, it is used to reduce the possibility of a stroke, which is caused when arteries in the brain are blocked by clotted blood. While (1), (2), and (4) may be positive uses of aspirin, they do not relate to its ability to retard clotting.

16. (2) As the north pole of one magnet is brought closer to the north pole of another magnet, the forces pushing them apart get stronger.

17. (2) The layer of ozone serves as a protective shield for life on earth by absorbing the X rays and ultraviolet waves that are emitted by the sun.

18. (4) The half-life of a substance is the period of time required for one-half of a given radioactive mass to disintegrate. If a radioactive material with a weight of 1,000 grams has a half-life of 10,000 years, then in 10,000 years there would only be 500 grams of the substance. In 20,000 years this would be reduced by half again, to 250 grams.

19. (5) Fossils are the remains of prehistoric plants and animals that have been preserved in rocks.

20. (2) The last paragraph states that light is a mixture of different wavelengths, ranging from violet, which has a relatively short wavelength, to deep red, which has a longer wavelength.

21. (5) The third paragraph mentions wavelengths ranging from 10 trillionths of a meter to 10,000 kilometers. From this you could conclude that the length of electromagnetic waves varies greatly.

22. (1) The distance labeled "19 meters" extends from one trough to the next. According to the third paragraph, this would measure the wavelength.

23. (2) Cosmic rays are not mentioned as part of the electromagnetic spectrum.

24. (3) The third paragraph mentions that white light can be broken down into its component colors, as in a rainbow. It is reasonable to conclude that the glass prism can break down light into its component colors.

25. (2) Look across the row that starts with blood type A. You can see that a person with blood type A can receive blood from blood types O and A.

26. (4) Look across the row on the table that starts with blood type AB. You can see that a person with blood type AB can act as a donor only to a person with blood type AB.

27. (1) Since people with blood type O can donate blood to people with all blood types, blood type O is often referred to as the universal donor.

28. (5) The last sentence of the first paragraph states that babies born to smokers are an average of six to seven ounces lighter at birth.

29. (5) Abruptio placentae is described in the third paragraph.

30. (1) In the first paragraph, the author mentions that the lack of oxygen may explain why babies who are born to smokers are an average of six to seven ounces lighter at birth than the babies of nonsmokers.

31. (3) This is the only choice not specifically mentioned. All of the other choices can be found in the second paragraph.

32. (2) A mother who smokes more than a pack a day increases the chances of her baby's dying by 20 percent. Those who smoke more than a pack a day increase those chances to 36 percent. Mothers who smoke more than a pack a day increase the risk of their baby's death by 16 percent.

33. (1) The fourth paragraph states that air is a poor conductor and insulates the cloud and ground charges. This prevents a flow of current until the electrical potential becomes great enough to overcome the resistance of the air.

34. (5) The passage states that the bottom of a storm cloud is dominated by negative charges, inducing a positive charge on the ground below. It also mentions that the attraction between negative and positive ions makes the ground current flow up buildings and trees. This allows you to conclude that the positive charges on the ground are built up because of their attraction to the negative charges in the cloud.

35. (3) The caption for the fourth diagram mentions that a typical cloud-to-ground stroke takes about one second.

36. (1) The fourth paragraph mentions that the attraction between the positive and negative charges makes the positive ground current flow up elevated objects. Answer (1) is the only one that does not include an object that is elevated above the surrounding area.

37. (2) Hybrid sparrows are not mentioned as a contributor to the extinction of the dusky. Rather it is the creation of hybrid sparrows that will preserve most of the dusky's characteristics.

38. (5) The creation of a hybrid sparrow will save most of the dusky sparrow's characteristics.

39. (3) While all of the titles include specific elements of this passage, only choice (3) represents the main idea.

40. (1) Since mosquito abatement procedures were, in part, responsible for the destruction of the dusky's habitat, you can infer that mosquitoes were the most likely food source.

41. (3) The passage does not mention other organisms being threatened with extinction. If the other seaside sparrows were threatened, then scientists would not attempt to use them to create a hybrid. Options (1) and (2) are unreasonable, since all of the transformations had to affect all of the species in the area.

42. (3) The last paragraph states that soap scum can be reduced or eliminated by rinsing soap out with acidic substances such as vinegar or lemon juice.

43. (4) Since shampoos for babies are usually amphoteric, and baby shampoos are usually mild, it is reasonable to assume that this is the mildest type of synthetic detergent.

44. (5) Synthetic detergents do not react with the calcium in water. Scum deposits do not form, even in the hardest water.

45. (5) The second paragraph states that the relatively high amounts of calcium in hard water combine with the soap to form soap scum.

46. (1) The passage states that when soap is used in hard water a gummy substance forms and that the familiar ring in the bathtub is composed of this substance, along with dirt and other matter. This is not true of the synthetic shampoos.

47. (2) The second paragraph states that an exposure of 500 rems over a brief period of time would be lethal in 50 percent of the cases.

48. (2) The fifth paragraph states that cancer is the most feared type of somatic damage.

49. (4) In spite of what one would think, the fourth paragraph states that cell damage is considered more harmful than cell death. Dead cells can be replaced, but damaged cells can replicate.

50. (2) The last two sentences state that the latency period—the time between exposure and effect—for genetic damage is a generation or more.

51. (1) While choices (2), (3), and (4) contain items mentioned in the passage, only choice (1) covers the main idea.

52. (4) The fifth paragraph states that when a reproductive cell is damaged a mutation can be caused, and the damage is transmitted to future generations.

53. (5) From the passage you can infer that it is difficult to predict the long-term effects of low levels of radiation because the latency period is so long.

54. (2) In May, the difference between the average temperatures for the two cities is less than one degree. The difference is greater for each of the other choices.

55. (3) The graph shows that Chicago's summer is hotter and its winter is colder. Therefore, its temperatures are more extreme than San Francisco's.

56. (4) The end of the second paragraph states that wind is created by the uneven heating of the earth's atmosphere by the sun.

57. (5) The second paragraph states that oil, gas, and coal all come from the fossils of plants that grew by converting sunlight to energy. These plants died thousand of years ago. Biomass consists of the same plant matter but is very recent.

58. (4) The energy the animals used in their work came from plants (oats, corn, barley, hay), which had used the sun's energy for growth.

59. (2) The passage states in the first paragraph that a drop of only 2 percent would start a glacial advance that would turn our earth into a giant snowball. While the temperature of the earth would drop, the drop would be much greater than a couple of degrees.

60. (1) The second paragraph states that geothermal energy is heat from the deep interior of the earth.

POST-TEST EVALUATION CHART

Below is a chart that you can use to evaluate your preparedness for the GED science test.

Circle the number of any questions that you missed. Then use the headings on the chart to see which areas you need to review. Read down the chart for content areas and across the chart for the reading and interpretation skills that are discussed in the first two chapters of this book. The heading "Prior Knowledge" refers to questions that require you to recall facts you may have previously learned or read.

The questions in darker type require pictoral or graphic interpretations skills. If you missed more than four of these eleven questions, review the chapter that starts on page 53.

	Main Idea	Finding Details	Application of Ideas	Inferences and Conclusions	Prior Knowledge
Biology	39, 51	***25***, ***26***, 28, 29, 30, 31, 37, 38, 47, 48, 49, 50, 52	7, 9, 15	***3***, ***8***, ***27***, 32, 40, 41, 53	1, 2, 4, 5, 6
Earth Science		33, ***35***, ***54***, 57, 60	36, 58	34, ***55***, 59	10, 17
Chemistry		42, 45	14, 18	43, 44, 46	13
Physics		20, ***22***, 23	***11***, 12, ***16***	21, 24	

Now count the number of circles. Subtract this number from 60. Then use the chart to determine what an equivalent GED score would be.

Scoring Chart

Number Correct	GED Standard Score
15 or fewer	35 or lower
20	40
26	45
33	50
40	55
46	60
51 or more	65 or higher

Answer Grid

1 ① ② ③ ④ ⑤

2 ① ② ③ ④ ⑤

3 ① ② ③ ④ ⑤

4 ① ② ③ ④ ⑤

5 ① ② ③ ④ ⑤

6 ① ② ③ ④ ⑤

7 ① ② ③ ④ ⑤

8 ① ② ③ ④ ⑤

9 ① ② ③ ④ ⑤

10 ① ② ③ ④ ⑤

11 ① ② ③ ④ ⑤

12 ① ② ③ ④ ⑤

13 ① ② ③ ④ ⑤

14 ① ② ③ ④ ⑤

15 ① ② ③ ④ ⑤

16 ① ② ③ ④ ⑤

17 ① ② ③ ④ ⑤

18 ① ② ③ ④ ⑤

19 ① ② ③ ④ ⑤

20 ① ② ③ ④ ⑤

21 ① ② ③ ④ ⑤

22 ① ② ③ ④ ⑤

23 ① ② ③ ④ ⑤

24 ① ② ③ ④ ⑤

25 ① ② ③ ④ ⑤

26 ① ② ③ ④ ⑤

27 ① ② ③ ④ ⑤

28 ① ② ③ ④ ⑤

29 ① ② ③ ④ ⑤

30 ① ② ③ ④ ⑤

31 ① ② ③ ④ ⑤

32 ① ② ③ ④ ⑤

33 ① ② ③ ④ ⑤

34 ① ② ③ ④ ⑤

35 ① ② ③ ④ ⑤

36 ① ② ③ ④ ⑤

37 ① ② ③ ④ ⑤

38 ① ② ③ ④ ⑤

39 ① ② ③ ④ ⑤

40 ① ② ③ ④ ⑤

41 ① ② ③ ④ ⑤

42 ① ② ③ ④ ⑤

43 ① ② ③ ④ ⑤

44 ① ② ③ ④ ⑤

45 ① ② ③ ④ ⑤

46 ① ② ③ ④ ⑤

47 ① ② ③ ④ ⑤

48 ① ② ③ ④ ⑤

49 ① ② ③ ④ ⑤

50 ① ② ③ ④ ⑤

51 ① ② ③ ④ ⑤

52 ① ② ③ ④ ⑤

53 ① ② ③ ④ ⑤

54 ① ② ③ ④ ⑤

55 ① ② ③ ④ ⑤

56 ① ② ③ ④ ⑤

57 ① ② ③ ④ ⑤

58 ① ② ③ ④ ⑤

59 ① ② ③ ④ ⑤

60 ① ② ③ ④ ⑤